Tapered Beams in MEMS

Wajih U. Syed • Ibrahim (Abe) M. Elfadel

Tapered Beams in MEMS

A Symbolic Modeling Framework with Applications to Energy Harvesting

Springer

Wajih U. Syed
Masimo (United States)
Irvine, CA, USA

Ibrahim (Abe) M. Elfadel
Khalifa University of Science
and Technology
Abu Dhabi, United Arab Emirates

ISBN 978-3-031-66390-1 ISBN 978-3-031-66391-8 (eBook)
https://doi.org/10.1007/978-3-031-66391-8

This Springer imprint is published by the registered company Springer Nature Switzerland AG
The registered company address is: Gewerbestrasse 11, 6330 Cham, Switzerland

To our beloved families
with our utmost gratitude.

كلّ فنّ لا يُفيدُ علماً لا يُعوّلُ عليه
محي الدين بن عربي

Foreword

In 2013, an ambitious collaboration unfolded across three distinct organizations: a semiconductor business in Singapore, a government research institute (also based in Singapore), and a non-profit post-graduate educational institution in Abu Dhabi, UAE. Their shared mission was to investigate the intricate world of microelectromechanical systems (MEMS) and push the boundaries of innovation toward commercialization. This trilateral partnership, meticulously described in its master agreement, operated under the auspices and financial support of the Abu Dhabi and Singaporean governments, a testament to the global significance of their endeavor. At the helm of this scientific voyage stood GlobalFoundries Singapore (GFS), entrusted with steering the research ship and managing the allocated funds. The other two key players were the Singapore Institute of Microelectronics (IME) and the Abu Dhabi Masdar Institute (MI). During the ensuing four years, the joint team fervently collaborated, focusing their complementary expertise on advancing MEMS technologies across four pivotal domains: ultrasound, motion sensing, energy harvesting, and optical MEMS. The joint team, consisting of approximately 30 principal investigators, worked tirelessly to materialize their ideas into working miniatured systems for emerging applications. Weekly conference calls for each project, led by IME project leaders and under the joint program leadership of GFS-IME, tracked progress and ensured alignment with the vision of the grantors. The semiannual meetings, hosted in Singapore by IME or GFS, provided a platform to report progress and exchange technical information. These gatherings convened a steering committee composed of senior managers from all three organizations: A snapshot of one such meeting graces the closing pages of this foreword. The division of contribution among the trio was clear: IME spearheaded the MEMS platform development, building the technical chain from design, integration flow, Process Design Kit (PDK) up to device integration for the four application domains. These included piezoelectric multiple ultrasound transducers (pMUTs), inertial motion units (IMUs), piezoelectric vibration energy harvesting, and opto-electromechanical sensing. Meanwhile, MI focused on designing the test vehicle according to the design rules, thus mimicking foundry customers while simultaneously nurturing MEMS education and research in Abu Dhabi. GFS, in turn, performed the technol-

Fig. 1 Group photo from one of the many MEMS Twinlab meetings in Singapore. This photo was taken on October 30, 2014

ogy transfer of the developed MEMS platforms to enhance their commercialization value. I extend my heartfelt appreciation to the authors for their remarkable efforts in bringing this book to fruition. It stands as a testament to the capacity building initiatives in MI, now an integral part of Khalifa University, and reflects the collaborative spirit that transcends geographical boundaries in Singapore and Abu Dhabi (Fig. 1). Moreover, I believe this book will serve as a valuable resource for researchers who are eager to immerse themselves in new technological challenges with the goal of transforming ambitious dreams into tangible working micro-devices.

Principal Scientist, A*STAR IME
Singapore, May 2024

Dr. Peter Chang

Preface

The promise of ubiquitous wireless sensors in support of the Internet of Things cannot be realized without a viable self-powering solution. Researchers have proposed piezoelectric vibrational energy harvesters as a promising technology for powering these autonomous electronic devices [1, 2]. Clamped-free beams with a tip-mass are widely known to perform well among piezoelectric vibrational energy harvesters. In such structures, however, the stress is concentrated at the fixed end, thus reducing the average stress while also risking early fatigue-induced device failure. Tapered beams, and in particular triangular beams, have been shown to distribute stress more evenly along the beam length [3] and therefore make a greater charge separation available for harvesting.

High fabrication costs make the microfabrication industry extremely sensitive to design failures. An iterative design scenario requires working with mathematical and computational models of the device. Numerical simulation tools for MEMS devices that use finite element analysis (FEA) have matured over time and are fast becoming industry standards to verify device performance against design specifications [4, 5]. Such simulations are now an integral part of the design flow of MEMS devices. Given a list of specifications, computational models for the device are prepared, which are simulated to confirm the device behavior with respect to the design specifications. The design problem can become hard to solve when the design space increases with multiple degrees of freedom, and the design objective becomes highly multidimensional. In such cases, an optimization script is necessary to iterate over the design space and find the best design while the models need to be as fast as possible. The development of accurate and computationally efficient models of popular MEMS devices is crucial to building and improving MEMS CAD modules. Where analytical solutions are possible, they are preferred over numerical FEA solutions for their computational efficiency and structure. Structured closed-form analytical solutions enable deep insight into model dynamics that helps to develop more powerful analysis techniques. Efficient solutions to analytical models of the tapered beam structure are hard to derive, as their direct solutions are expressed in terms of special functions such as Meijer G and hypergeometric functions [6]. In this work, we have used the perturbation theory [7] to derive modal, steady-state, and transient solutions to the tapered beam piezoelectric energy harvester (TPEH) governing system by considering the beam taper as a perturbation around a nominal rectangular case.

The steps of the microfabrication process, such as various types of deposition, lithography, and etching, take the desired geometrical parameters as input and are designed to produce a device geometry according to the design specifications. The specifications are fed into the process steps as nominal values, but each of these steps has intrinsic variability that causes a tolerance spread of the features around their nominal values. These variabilities are functions of the apparatus that performs these steps and the underlying physics. The manufacturing variability thus explained gives rise to randomness in the geometry and material properties of the manufactured device. These uncertainties adversely affect the performance of the MEMS devices and also propagate to the circuit level to influence the system behavior. Thus, a complete system-level uncertainty model has to include the variabilities at both the device and the circuit levels and the propagation of these uncertainties through both the device and circuit models.

A typical design flow employed in the industry works with deterministic models and does not consider the effect of uncertainties on device performance. The target nominal behavior would reflect in the average devices thus manufactured, but the lack of uncertainty models results in various challenges for device development. The device yield from the microfabrication process can only be reliably predicted if a sufficiently accurate model exists for the uncertainties influencing the device performance. Device performance variability helps measure the robustness of the device. When the design process is agnostic of the process variabilities, it cannot help optimize on minimizing the influence of the uncertainties on the device performance. Also, without an uncertainty model, the error between the observed device performance and the model output cannot be well characterized, and hence any attempts of model validation will have to rely upon the allowed error tolerances derived from certain rules of thumb.

The primary UQ strategy in use is the non-intrusive methodology of performing MC simulations on the device models. These device models are often FEA models, making MC analysis a very slow process. The speed of MC-based UQ is improved by using the semi-analytical TPEH model developed in this work. UQ techniques based on generalized polynomial chaos (gPC) expansion have been proven to be much faster than MC methods, though they are intrusive and require more effort to setup. The tapered PEH model developed in this book helps to develop the intrusive gPC-based UQ algorithms. It also enables a more comprehensive comparison between non-intrusive and intrusive UQ strategies for the tapered PEH.

There are two major parts to this book. In the first part, a complete and working model of a tapered beam unimorph PEH is developed. This model is capable of modal analysis, steady-state analysis, and transient analysis of a TPEH under various loading conditions. Toward the ultimate goal of developing a complete distributed parameter tapered PEH model, the book takes the following steps:

1. As a precursor to developing the electromechanical model, the vibrational dynamics of a multi-layered, tapered, prismatic, micro-cantilever structure is established using perturbation theory where perturbation is applied around the rectangular beam with the taper ratio as the perturbation parameter. The result is a novel semi-analytical approach of computing the resonance frequencies and mode shapes of vibrations of a tapered micro-cantilever structure.
2. Using the perturbation-modal analysis performed on tapered cantilevers, a full electromechanical model of the tapered PEH is developed. A state-space model of the tapered PEH thus prepared is then solved using state-space solvers.
3. An equivalent circuit model (ECM) of the tapered PEH is also developed. This circuit model is capable of simulations under various electrical loading conditions.
4. All proposed models are meticulously tested against FEA and circuit simulation tools.

In the second part of the book, various uncertainty quantification schemes are explored for the tapered PEH. The highlights of this second part are the following:

1. We propose an original uncertainty propagation graph of the tapered PEH model.
2. We use the semi-analytical model of the tapered PEH developed in the first part of the book to propose an efficient non-intrusive UQ based on the Monte Carlo method. We also take advantage of probabilistic models to propagate uncertainty through the tapered PEH model using the uncertainty graph.
3. We further use the semi-analytical model to propose intrusive UQ methods that are faster than Monte Carlo. A polynomial chaos expansion of the tapered-beam EH model is performed via the collocation approach, culminating in an augmented stochastic system.
4. Finally, we formulate a variability-aware MEMS design flow based on the TPEH model and the proposed UQ methods.

One unique feature of the proposed methodology and models is that they are symbolic. At every step of the modeling process, the dependence of the mathematical expressions and formulae on geometric and material properties is explicit. This explicit dependence has the distinct advantage of providing physical insight, tracking the root causes of inaccuracies, and clarifying the roles of parametric uncertainties. The price to pay for such explicit dependence is complex formulations, intricate expressions, and notational inflation. What has made the endeavor manageable for us is the symbolic codebase that we have developed using Mathematica. Other computer algebra systems could have been used, but our choice of Mathematica was mainly driven by familiarity and background. We have kept the Mathematica codebase flexible and reusable so as to adapt it to other physical domains where taper is a geometric design parameter.

Now, we will take a brief tour of the book chapters. Chapter 1 introduces the piezoelectric MEMS energy harvesters that the book uses to illustrate the symbolic modeling methodology. It describes the motivation, reviews the relevant literature, and provides the background information needed in the subsequent chapters. Chapter 2 introduces the Euler-Bernoulli beam theory, which lays the foundation of the mathematical model upon which the symbolic methodology is built. Chapter 3 discusses the importance of the tapered shape for the design and fabrication of energy harvesting devices and also presents the design, fabrication, and characterization work on tapered PEH devices. Chapter 4 discusses the formulation of a semi-analytical model using perturbation analysis to solve for the vibrational dynamics of a tapered cantilever in terms of its mode shapes. Chapter 5 describes the symbolic library developed and used in building the perturbation-based model. Chapter 6 builds on the tapered beam model and completes the coupled electromechanical model for a tapered beam piezoelectric energy harvester with distributed parameters before suggesting its equivalent circuit model. Chapter 7 sets the benchmark analysis for uncertainty quantification using the Monte Carlo simulation approach and presents the results for Monte Carlo uncertainty quantification of a tapered beam energy harvester. Chapter 8 presents the polynomial chaos expansion of the tapered beam energy harvester and then explains the different methods to solve the uncertainty system. Chapter 9 combines the TPEH model with UQ and proposes a variability-aware design flow that is based on the UQ of analytical models. other physical domains and points to symbolic modeling directions that can take advantage of the methodology of this book.

Irvine, CA, USA
Abu Dhabi, UAE
April 2024

Wajih U. Syed
Ibrahim (Abe) M. Elfadel

References

1. Shad Roundy et al. "Improving power output for vibration-based energy scavengers". In: *IEEE Pervasive computing* 4.1 (2005), pp. 28–36.
2. Shad Roundy. "On the effectiveness of vibration-based energy harvesting". In: *Journal of intelligent material systems and structures* 16.10 (2005), pp. 809–823.
3. Jessy Baker, Shad Roundy, and Paul Wright. "Alternative geometries for increasing power density in vibration energy scavenging for wireless sensor networks". In: *3rd international energy conversion engineering conference*. 2005, p. 5617.
4. *CoventorWare Field Solver Reference*. Coventor Inc. 2016.
5. Coventor Inc. *MEMS+ Users Guide*, 2002.
6. "Han-Chung" Wang. "Generalized hypergeometric function solutions on the transverse vibration of a class of nonuniform beams". In: *Journal of Applied Mechanics* (1967), pp. 702–708.
7. Steven A Orszag and CM Bender. "Advanced mathematical methods for scientists and engineers". In: *Mac Graw Hill* (1978).

Acknowledgements

This book is an outgrowth of the PhD thesis of the first author conducted under the supervision of the second author. The thesis work was completed at the Masdar Institute (MI), now part of Khalifa University, Abu Dhabi, United Arab Emirates, between 2013 and 2017, in the framework of the MEMS TwinLab partnership between the Institute of Microelectronics (IME) in Singapore and MI. The research on which the book is based would not have been possible without the help and support of so many dedicated colleagues across three continents and several organizations.

First and foremost, we acknowledge the help of the MEMS Twinlab Energy Harvesting team, including Alabi Bojesomo and Owais Waheed from MI and Dr. Chengliang Sun from IME. The authors also thank Dr. Luca Daniel from MIT and Dr. Daniel Choi from MI for serving on the PhD thesis committee of the first author and sharing their insightful comments throughout the thesis research. In addition, we acknowledge the MI CAD team of MEMS Twinlab, including Drs. T. J. Zhang, Mohamed Ali, and Weilin Yang for their contributions to various aspects of computational modeling of MEMS devices.

The authors are grateful to Dr. Peter Chang from IME for his technical leadership and the deep insights he continuously shared on all aspects of MEMS design and technology across all Twinlab projects. In terms of our work schedule and organization, we greatly benefited from the management leadership of a number of very capable engineers at GLOBALFOUNDRIES Singapore (GFS) to whom day-to-day management was entrusted by the project sponsors. They included Siddharth Chakravarty, Aveek Chatterjee, Dr. Zouhair Sbiaa, and Dr. Nagarajan Ranganathan.

MEMS Twinlab was a huge opportunity for the UAE to build research and education capacity in the area of MEMS technology. MI stepped up to this opportunity and established a state-of-the-art infrastructure of experimental, computational, and curricular facilities in support of the Twinlab projects. We would like to acknowledge the following MI faculty who contributed to building this infrastructure almost from scratch along with their staff and students: Drs. Mohamed Ali, Daniel Choi, Marcus Dahlem, Clara Dimas, Anatoly Khilo, Mahmoud Rasras, Irfan Saadat, Firas Sammoura, Jaime Viegas, Jerald Yoo, and T. J. Zhang. The MI graduates of the MEMS Twinlab projects are now involved in professional careers with world-class organizations in Europe, Asia, and North America. We are proud of their contributions and achievements.

One major contributor to the enablement of MEMS design at MI was Coventor, now part of Lam Research. In particular, we thank Dr. Chris Welham for his leading role in providing the MI researchers with training and support on Coventor's tools and workflows.

A team of senior leaders, including Dr. Alex Gu from IME, Dr. Rakesh Kumar from GFS, and Dr. Steve Griffiths from MI, provided overall guidance and direction to MEMS TwinLab for a four-year period. We thank them for the empowering environment they created and the collaborative partnerships they nurtured.

Warm thanks are also due to our sponsors: the Mubadala Investment Company in the UAE, represented by Dr. Rafic Makki, and the Singapore Economic Development Board represented by Lee Ming Xuan, for their generous support and visionary leadership.

Finally, we thank Springer's Charles Glasser and his editorial team for their guidance and the almost infinite patience with which they have handled our crazy writing and editing schedule.

Contents

Nomenclature

Abbreviations

AlN	Aluminum Nitride
CAD	Computer aided design
CDF	Cumulative Density Function
CW	CoventorWare - An FEA software by Coventor Inc. for analyzing MEMS.
DAE	Differential Algebraic Equation
ECM	Equivalent Circuit Model
EH	Energy Harvester
FEA	Finite Element Analysis
FEM	Finite Element Method
FEM	Finite element method
GF	GLOBALFOUNDRIES
gPC	Generalized Polynomial Chaos
IME	Institute of Microelectronics
IMU	Inertial measurement unit
ITPS	Iterative Perturbed Tapered System
KDE	Kernel Density Estimation
KNN	Potassium Sodium Niobate
MC	Monte Carlo
MEMS	Microelectromechanical systems
MEMS+	A fast MEMS FEA tool created by Coventor Inc.
MI	Masdar Institute
PCE	Polynomial Chaos Expansion
PDF	Probability Density Function
PEH	Piezoelectric Energy Harvester
PZE	Piezoelectric
PZT	Lead Zirconate Titanate
Q factor	Quality factor
SDOF	Single Degree of Freedom
SEM	Scanning electron microscope

SFD	Squeezed-film damping	
SOI	Silicon on Insulator	
TC	Tapered cantilever	
TPEH	Tapered Piezoelectric Energy Harvester	
UP	Uncertainty Propagation	
UQ	Uncertainty Quantification	

Subscripts and Superscripts

E	Superscript E implies the quantity is evaluated at constant electric field	
i in $[\,]_i$	Layer index	
i, j, k, l	Generic indices for tensor or summation indexing	
N_p in $[\hat{\,}]_{r,N_p}$	Perturbation order used for approximation	
r	Mode number	
S	Superscript S implies the quantity is evaluated at constant strain	
T	Superscript T implies the quantity is evaluated at constant stress	

Mathematical Symbols

α	Taper ratio	[dimensionless]
$\bar{\kappa}_r$	Normalized average curvature of the mode shapes	
β_r	Eigenvalue for mode r	
$\boldsymbol{D}$	Electric displacement tensor	$[Cm^{-2}]$
$\boldsymbol{E}$	Electric field tensor	$[Vm^{-1}\ or\ NC^{-1}]$
$\boldsymbol{S}$	Strain tensor	[dimensionless]
$\boldsymbol{s}$	Elastic compliance matrix	$[m^2N^{-1}\ or\ Pa^{-1}]$
$\boldsymbol{T}$	Stress tensor	$[Nm^{-2}\ or\ Pa]$
χ_r	Piezoelectric coupling coefficient of mode r	
$\delta(x)$	Dirac delta function	
$\eta_r(t)$	Modal response function of mode r	$[m]$
$\hat{\beta}_{r,N_p}$	Tapered cantilever eigenvalue estimate for mode r and perturbation order N_p	
$\hat{\phi}_{r,N_p}(x)$	Tapered cantilever mode-shape estimate for mode r and perturbation order N_p	
λ_r	rth eigen-number of the beam.	
$\mu_{r,n}, \mu_n$	The nth perturbation component of the eigenvalue of the mode-shape system for mode r	
ω_r	Natural frequency of mode r	$[rad\,s^{-1}]$
$\phi_r(x)$	Mode shape function of mode r evaluated at position x	
$\psi_{r,n}, \psi_n$	The nth perturbation component of the beam mode shape for mode r	
$\theta(x)$	Piezoelectric coupling factor	$[NmV^{-1}\ or\ C]$
ε	Permittivity	$[Fm^{-1}\ or\ NV^{-2}]$
ζ_r	Damping ratio of mode r	
b_0	Beam width at the anchor i	$[m]$
C_r	Equivalent mechanical capacitance for mode r	

c_s	strain rate damping coefficient	$[kgs^{-1}]$
c_a	viscous air damping coefficient	$[kgs^{-1}]$
d	Piezoelectric strain coefficient matrix	$[mV^{-1}]$
$g(t)$	Translational transverse displacement at time t	$[m]$
$h(t)$	Angular displacement about y-axis at time t	$[rad]$
I	Moment of inertia of the beam	$[kgm^2]$
$i_r(t)$	Current in rth branch of the equivalent circuit	
L	Beam length i	$[m]$
l	Length of the beam	$[m]$
L_r	Equivalent mechanical inductance for mode r	
$m(x)$	Mass per unit length of the beam at x	$[kgm^{-1}]$
$M(x,t)$	Total internal bending moment at x and t	$[Nm]$
N_l	Number of layers of the cantilever structure	
N_r	Ideal transformer turns ratio modeling electromechanical coupling for mode r	
N_x	Number of spatial samples	
$P_k(f)$	kth moment of $f(x)$	
$P_k^{(m)}(f)$	kth mass-weighted moment of $f(x)$	
$q_r(t)$	Charge collected for mode r	
R_r	Equivalent mechanical resistance for mode r	
$S(z,x,t)$	Strain distribution	
$T(x,t)$	Stress distribution	$[Nm^{-2}]$
t_i	Thickness of layer i	$[m]$
$v(t)$	Voltage at EH terminals	$[V]$
$V_r(t)$	Equivalent voltage of mechanical actuation for mode r	
w_b	Base motion of the device	$[m]$
w_{rel}	Deflection of the beam neutral axis in z-axis	$[m]$
YI	Flexural rigidity of the beam	$[mkg^2s^{-2}]$
M(x,t)	Total internal bending moment at x and t	$[Nm]$

Chapter 1
Introduction to Piezoelectric MEMS Energy Harvesting

The book discusses a complete working model of a tapered cantilever piezoelectric (PZE) energy harvester (EH) which is used to build upon and discuss the uncertainty quantification strategies for variability-aware design of these devices. This chapter provides background information for the modeling work presented in the forthcoming chapters. The first part of the chapter introduces MEMS piezoelectric vibration energy harvesting and the physics involved, along with a survey of PZE devices. The second part of this chapter provides an introduction to uncertainty quantification and variability-aware design.

1.1 MEMS Piezoelectric Vibrational Energy Harvesting

Among the many ambient energy sources available in our environment for harvesting, vibration energy has recently been recognized as a readily available and easily harvestable source of energy. Energy harvesting implies the conversion of ambient energy to electrical energy and making it available to power devices and electronics. Vibration energy can be harvested using various energy conversion mechanisms including electrostatic, electromagnetic, and piezoelectric transduction. Piezoelectric vibration energy harvesters are the subject of this book due to the benefits they exhibit in the context of microelectromechanical systems (MEMS).

1.1.1 Microelectromechanical Systems

MEMS are micrometer-scale mechanical systems that are built using silicon-based fabrication processes borrowed from CMOS foundries. They use various material properties to sense, actuate, or transduce signals and energy. MEMS sensors and

W. U. Syed, I. (Abe) M. Elfadel, *Tapered Beams in MEMS*,
https://doi.org/10.1007/978-3-031-66391-8_1

actuators have quickly become ubiquitous and are now necessary components of most high-tech products around us. They are as important to smartphones and tablets as transistors were to consumer electronics in the 1960s. Today, a smartphone contains up to 10 MEMS devices. The Yole Development market forecast places the MEMS industry close to an 18B USD market by 2026 [1].

1.1.2 Energy Harvesting

The Internet of Things has become a reality today and has spread quickly to all the technical solutions used by society. The vision of ubiquitous wireless sensors has great potential in areas such as building monitoring, environmental control, wearable electronics, and structural monitoring of machines in military and commercial applications. A *network of everything* requires sensors scattered in diverse remote locations including many that are not easily accessible. A key problem in making this vision a reality is how to effectively deliver power to these devices. Conventional batteries have a limited lifespan and thus incur regular replacement overhead and also have a hazardous impact on the environment. In applications such as structural health monitoring in remote locations and other geographically inaccessible remote sensors, battery charging and replacement can be expensive, cumbersome, and often impossible. Moreover, in wearable electronic applications, users would prefer to avoid having to recharge or replace batteries.

As advances in low-power microelectronics [2–4] have reduced sensor power consumption to save battery cost, the lower power required has opened doors to energy harvesting solutions on-chip that scavenge ambient energy from the environment. Such solutions have far less impact on the environment compared to fossil fuels and electrochemical batteries and can also help eliminate the need for batteries and tentatively extend the lifetime of the device indefinitely.

Among the potential energy harvesting solutions that can meet the energy needs of microelectronic devices are photovoltaic, thermoelectric, and kinetic approaches. Solar power systems are the most popular and widely used energy harvesting solution. Although they have their advantages [5], their low conversion efficiencies and high cost, as well as the intermittent availability of sunlight, cause challenges in their adoption as the first choice strategy for powering microsystems. *Thermoelectric generators* have also attracted considerable attention as a viable power supply strategy for microsystems, but their limited energy conversion efficiencies and operational temperatures have restricted their usage. More recently, work on energy harvesting from external mechanical excitation has gained momentum. *Kinetic generators* convert mechanical energy into electrical energy by transducing mechanical displacements into electrical voltage. Figure 1.1 shows a power density *versus* voltage plot for common regenerative power solutions. This plot indicates that the power density values of piezoelectric energy harvesting solutions overlap with those of lithium-ion batteries implying that piezoelectric energy harvesting is a viable strategy for powering microsystems.

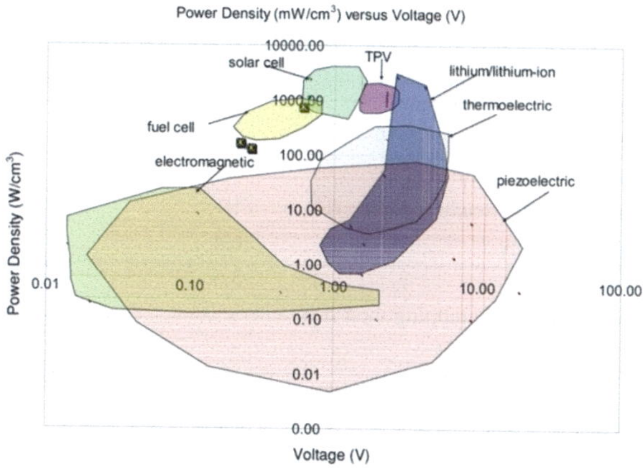

Fig. 1.1 Power density *versus* voltage for common power supply strategies [5]

1.1.3 Vibrations as a Source of Energy

Mechanical vibration is an ever-present source of energy in our environment. From the rustling leaves to the hum you hear from the air-conditioning ducts, there are vibrations all around us. All objects resonate at frequencies selected by the properties of their structures. Vibration characteristics from common sources in factories, workspaces, vehicles, and houses have been investigated by Roundy et. al. [6, 7], and their results are shown in Table 1.1. The study indicates that most sources in our environment resonate with low peak frequencies (< 200 Hz) but with a relatively flat acceleration spectrum in the frequency range below 1 kHz. Other sources such as vehicle tires and aircraft engines exhibit vibrations at higher frequencies reaching 1 kHz. Information about vibration sources is important to consider when designing an energy harvester. The reported data indicates that an energy harvester designed to scavenge energy from one of the reported sources should be designed for a resonance frequency of less than 1 kHz.

Vibrational energy harvesting offers a promising solution for structural health monitoring and wearable electronics among other applications. The devices developed as part of this work target remote tire pressure monitoring, among other applications. Wearable electronics is a growing trend that is accelerated by the inundation of smartwatches and fitness bands. A growing elderly population and an increasing awareness of health and fitness lead to an ever-growing demand for wearable health products. As shown in Fig. 1.2, Yole Development has predicted that wearables will increase in demand and market value, with consumer and healthcare products, in particular, seeing a significant increase in demand over the next five years [1]. A continuous supply of energy is crucial for making wearable electronics sustainable and an energy harvesting solution will increase market acceptance of wearables [13]. A patch-type healthcare sensor in combination with a chest band for monitoring health without expensive batteries and Ag/AgCl electrodes is reported to consume only 12 μW of power [14, 15]. Yole Development's report [16] predicts

Table 1.1 Characterizing some vibration sources [6–10]

Vibration source	Acceleration (m/s^2)	Resonance frequency (Hz)
Car engine compartment	12	200
Car hood (750 rpm)	0.0744	35.6
Car hood (3000 rpm)	0.257	147.5
Base of 3-axis machine tool	10	70
Blender casing	6.4	121
Clothes dryer	3.5	121
Person nervously tapping their heel	3	1
Car instrument panel	3	13
Door frame just after door closes	3	125
Small microwave oven	2.5	121
HVAC vents in office building	0.2–1.5	60
Windows next to a busy road	0.7	100
CD on notebook computer	0.6	75
Second story floor of busy office	0.2	100
Bridge railing	0.0215	171.3
Medium tree	0.000985	16.3
Small tree	0.003	30
Railway	1.078–1.568	12–16
Truck	1.96–3.43	8–15
Ship	0.98–2.45	12–13
Vehicle tyre [11]	Not stated	40–1100
Helicopter hover [12]	Not stated	60.45

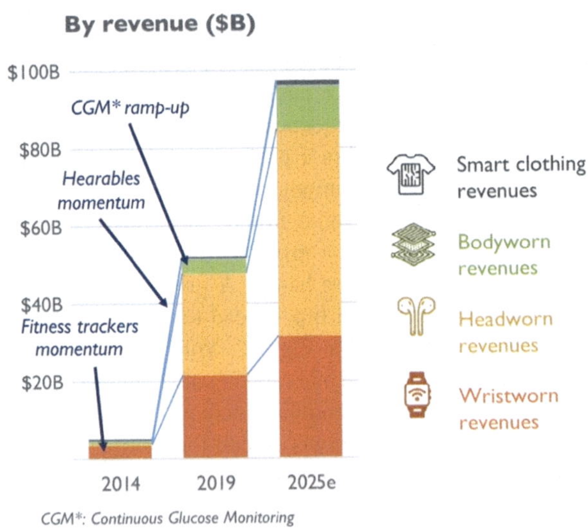

Fig. 1.2 Wearable market evolution with application breakdown between 2014 and 2025 [1]

that wearables will add 100 billion dollars to the consumer and medical electronic market.

Some recent advances in low-power consumption circuitry have enabled the development of integrated circuits that work with ultralow power of the order of nW. With rectification and signal conditioning circuits that can operate while consuming power on the order of nW to few μW, the use of vibration energy harvesters looks feasible.

1.1.4 Piezoelectric Transduction

Piezoelectricity is the phenomenon exhibited by some materials where an applied mechanical strain causes charge separation, and conversely, the application of an electric field causes material deformation. Piezoelectric materials are crystalline materials with nonuniform charge distribution within the crystal's unit cells, such that upon deformation the average positive and negative charge movements are different, hence causing a nonzero net electric polarization in the crystal. The principle of piezoelectric energy harvesters (PEHs) follows from this principle, where a spring structure ensures persistent motion under external excitation and thus continuing deformation of the structure that causes charge separation leading to electric potential at the electrodes available for harvesting. This piezoelectric phenomenon can be used in energy harvesters in one of two modes. In the parallel (33) mode, the electric potential harvested is parallel to the deformation, whereas, in the more commonly used transverse (31) mode, the polarization occurs in the 3-direction when deformation occurs in the 1-direction, or in other words, the electric potential is collected in a direction perpendicular to the deformation.

The piezoelectric materials most widely used include PZT and its various versions such as PZT-5H and PZT-5A [5], ZnO, and AlN. Each piezoelectric material can be characterized by a set of parameters including the strain coefficients and voltage coefficients that relate generated voltage to input stress [17]. A high energy density piezoelectric material has a large product of the strain coefficient and the voltage constant [18]. The piezoelectric properties of the material combine with its mechanical and electrical properties to yield an effective coupling factor that determines the converted energy and the harvester efficiency.

AlN thin films exhibit piezoelectric coupling factors comparable to PZT thin films [19] and some bulk piezoelectric materials and have been proven to be good enough for piezoelectric actuators, sensors, and energy harvesters [20].

1.1.5 Figures of Merit of Vibration Energy Harvesters

The designer of a piezoelectric MEMS energy harvester is seeking to maximize the total power generated across an applied load. This power, P, is related to the voltage developed at the terminals as

$$P = \frac{V^2}{Z_l},$$

where Z_l is the load impedance and V is the voltage that develops at the terminals as a result of the separation of charges by the piezoelectric layer. To compare different energy harvesters of varying geometry, the figure of merit often used is the power density, $\bar{P}$, which is defined as the power generated per unit volume. Since the thickness of MEMS devices is mostly uniform, a measure of the generated power per unit *area* often suffices. The power density, $\bar{P}$, therefore, often refers to the power generated per unit area instead. When comparing power densities of different devices, one must make sure that they are being compared when operating at the same input acceleration. A ratio of power density and the actuation acceleration can be very useful when comparing different energy harvesters.

Energy losses due to various damping mechanisms result in the deterioration of MEMS performance. The minimization of such losses is a key design concern. The quality factor, Q, is defined as the ratio of energy stored to energy dissipated and is usually used to quantify device damping due to energy losses. Where multiple sources of damping exist, the net Q factor, Q_{net}, can be evaluated using the reciprocal sum of the individual Q_i factors according to

$$\frac{1}{Q_{net}} = \sum_i \frac{1}{Q_i}.$$

Microstructures experience internal material damping of the mechanical structures such as thermoelastic damping (TED), anchor loss damping that results in energy dissipation into the substrate, and damping related to the fluid–structure interaction resulting in energy dissipation in the medium surrounding the device[21]. The Q-factor is related to the damping coefficient c of the device as

$$Q = \frac{m_0 \omega_0}{c},$$

where the effective mass m_0 and the resonant frequency ω_0 are obtained from the modal analysis of the device structure. Q factors are also related to the operating 2-dB bandwidth of the device as

$$Q = \frac{\omega_r}{\Delta\omega},$$

where ω_r is the resonance frequency of the structure and $\Delta\omega$ is the 2-dB bandwidth of the device. Where the design target is a high-bandwidth device, the figure of merit considered is the bandwidth, $\Delta\omega$, of the device, which is inversely proportional to the Q factor for a fixed resonant frequency.

Power generation per unit area can be maximized [6] by optimizing the size of the proof mass in relation to the device volume, maximizing the coupling coefficient [22] and the Q factor, and minimizing the impedance mismatch between the EH device and the loading circuit. The generalized electromechanical coupling is dependent on the piezoelectric coupling coefficients of the deposited AlN layer and the electromagnetic energy that it dissipates quantified by its *dielectric* loss. These material parameters are optimized in the MEMS foundry. Energy loss due to damping of the MEMS vibrating structure is represented quantitatively by the Q factor.

1.1.6 Survey of Piezoelectric Energy Harvesters

Though Williams and Yates [23] worked on vibration energy harvesters that used electromagnetic transduction, they concluded that the device was likely to perform poorly at low frequencies since its output power was proportional to the cube of the vibration frequency. Some other early work by Goldfarb and Jones [24] on power generation by stacked piezoelectric material identified storage challenges in the technology where the stored energy is returned to the excitation source, particularly in the presence of a capacitive load. Roundy [6]compared the transmission and coupling coefficients of electromagnetic, electrostatic, and piezoelectric transduction methods and deduced that the piezoelectric generators produce high voltages but low currents which scale down with the device size. The simplicity of devices based on piezoelectric transduction and their ease of integration with CMOS encouraged further research. The rapid growth of energy harvesting research has resulted in improving the power generation and transduction efficiency of EH devices and in meeting the various challenges identified. Sodano's review [25] captures some of this history that Anton builds on in his review [26] on the same subject. Table 1.2 shows a brief review of piezoelectric EH performance over recent years. In the table a_{ec} is the excitation acceleration, f_0 is the resonance frequency, V is the active volume, P is the power, and $\bar{P}$ is the power density.

1.2 Variability-Aware CAD of MEMS

To meet the promise of sensor ubiquity in the IoT era, MEMS design, much like CMOS design, must leverage CAD tools and methodologies to minimize sensor area, maximize yield, and reduce time to market. Most of the existing computational tools for MEMS design take the traditional MEMS design perspective of one MEMS process for one MEMS device and one sensor product. While very capable of modeling the 3D multi-domain physics of a MEMS sensor, these tools still require a considerable amount of manual tuning to adapt them to MEMS process changes, the variety of sensor environments, and the changes in product requirements. To

Table 1.2 Summary of piezoelectric energy harvesters [8, 27]

Reference	Material	a_{ec} (g)	f_0 (Hz)	V (mm^3)	P (μW)	$\bar{P}$ ($\frac{\mu W}{mm^3}$)	$\frac{\bar{P}}{a_{ec}^2}$ ($\frac{\mu W}{g\, mm^3}$)
Durou 2010	PZT	0.2	76	1.89	13.9	7.35	183.75
Kymisses 1998	PVDF	N/A	0.9	16,000	1300	0.08125	N/A
Umeda 1997	Bronze/PZT	N/A	N/A	214	19	0.08878	N/A
Shenk 2001	PZT dimorph	N/A	0.9	25,000	8400	0.336	N/A
White 2001	Steel/screen printed PZT-5H	0.234	80.1	125	2.1	0.0168	0.307
Roundy 2003	Brass/PZT/tungsten	0.255	120	1000	210	0.21	3.229
Roundy 2003	Brass/PZT/tungsten	0.255	120	1000	375	0.375	5.767
Sodano 2004	Mide Technology QP40 N transducer	N/A	30	240	11.9	0.04958	N/A
Duggirala 2004	^{63}Ni radioisotope silicon/PZT	N/A	35	60	0.001	0.000016	N/A
Bayrashev 2004	Terfenol-D/PZT/Terfenol-D composite	N/A	5	2185	80	0.03661	N/A
Marzencki 1969	AlN	1	900	2	0.1	0.05	0.05
Marzencki 1969	PZT	1	900	0.9	0.6	0.667	0.667
Jeon 1970	Micromachined silicon with IDT electrodes	10.8	13.9	0.027	1	37.04	0.318
Fang 2006	PZT, d_{31}	1	608	0.78	2.16	2.77	2.77
Hirasawa 2010	AlN	1	857	1.63	0.18	0.110	0.11
Muralt 2009	PZT, d_{33}	2	870	0.48	1.4	7.78	0.73
Morimoto 2010	PZT, d_{31}	0.5	126	0.26	5.3	20.5	82
Hajati 2010	PZT, d_{33}	4	1300	0.02	22	1100.0	68.75
Park 2010	PZT, d_{33}	0.39	528	1.05	1.1	1.05	6.90
Defosseux 2011	AlN, d_{31}	0.275	214	2.8	0.63	0.23	3.04
Xu 2011	PZT	1	329	20.9	7.35	0.35	0.35
Lei 2011	PZT	1	235	18.6	14	0.75	0.75
Aktakka 2011	PZT, d_{31}	1.5	154	27	205	7.59	3.37
Kanno 2012	KNN, d_{31}	1	1036	0.168	1.1	6.54	6.54

facilitate design tunability for the case of tapered beam EH devices, we formulate analytical models of the devices and extract their lumped models for fast analyses.

Advanced process techniques have enabled MEM devices to shrink feature sizes to meet industry demands. Unfortunately, even the most refined process technologies have precision limitations and inevitably cause unexpected deviations in the device dimensions from their nominal design values. Smaller feature sizes also render the small variations relatively more prominent. The increased variability in manufactured devices requires additional margins in design targets to satisfy yield goals. These margins are a function of the variability distribution as it affects critical device output and characteristics that define its performance. The CAD flow must also include an uncertainty quantification component that establishes margins for critical device design parameters.

The MEMS industry, therefore, seeks an automated methodology based on CAD for process- and variability-aware MEMS device optimization.

1.2.1 Variability in MEMS

The device parameters and consequently the device performance vary systematically with the deviation in nominal device dimensions due to the inevitable process variations that occur in the fabrication phase of the microsystems. MEMS fabrication processes such as plasma-enhanced chemical vapor deposition (PECVD), lithography, reactive ion etch (LRIE), isotropic etch, deep reactive ion etch (DRIE), and sputter metallization all exhibit different levels of variability owing to extraneous factors that affect how these processes are performed. Variation in temperature and pressure conditions of the process environment causes the process to execute differently each time. Process equipment is sensitive to its treatment before use and may cause wafer-to-wafer variations based on its thermal stabilization conditions. Wafer edge effects, process chamber asymmetry, and gas flow effects cause more thermally induced on-wafer variations. In addition, on-wafer variations are induced as surface stresses disturb the wafer planarity during processing. The distribution of process gasses while applying photoresist and the reactant diffusion during wet etching are also not exact and disturb on-wafer uniformity [28].

These limitations in process techniques then lead to inaccuracies in etching rates and etch depths, mask alignment and lithographic treatment, deposition rates and thicknesses, and sputtering uniformity leading to tolerances in design geometry. The design parameters are then affected by the tolerances in design geometry and their distributions. These distributions should ideally be inferred from the physics of the process steps that cause them. Owing to the lack of process models, these distributions are estimated using measurements on test structures from a history of fabrication using the foundry platform. They are mostly modeled as uniform distributions within a set of process step tolerances.

1.2.1.1 Types of Variability

Figure 1.3 shows the different sources of variabilities experienced when modeling MEM devices [29]. The *parametric uncertainty* is the uncertainty in the inputs to the model. These inputs could also appear as model parameters and should correspond to physical quantities. In terms of microfabrication, the parametric uncertainty shows up as process tolerances in device geometry as well as material variability. The lack of detail in the model of the device, be it a result of the lack of depth of the physics captured, inaccurate mathematical description of the physics,

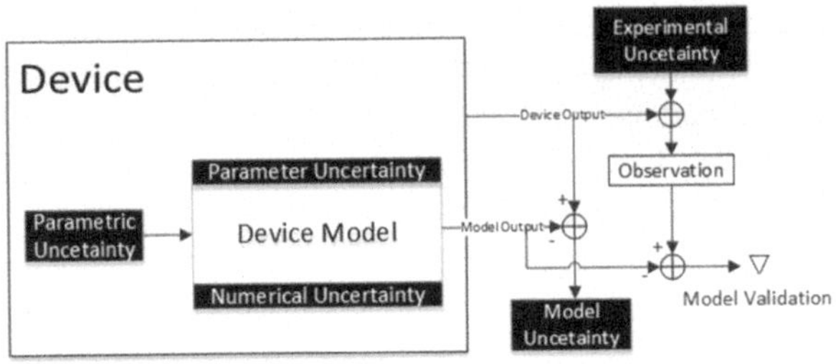

Fig. 1.3 Sources of uncertainty

or the simplifying assumptions in the interest of a more compact model, leads to uncertainty in model behavior. Such uncertainty, which is the *model or structural uncertainty*, explains the error between the actual device and the physics captured by the device model. Model parameters are often assigned empirical values owing to a lack of knowledge of some model details. The empirical values are subject to measurement errors and lead to inaccuracies modeled as *parameter uncertainty*. *Numerical uncertainty* includes quantization and sampling errors and is a function of the numerical algorithms employed. The algorithm itself is often parameterized to grid sizes, time steps, and truncation conditions and in that, numerical uncertainty influences the model in similar ways as parameter uncertainty. *Experimental uncertainty* is caused by the measurement error in observing the device output. Experimental uncertainty further constricts the model validation capacity of the experimental setup.

The uncertainty of the above sources in the context of MEMS reliability is related to the lack of understanding of the underlying models (*epistemic uncertainty*) or to the variability in the fabrication processes and layout (*aleatoric uncertainty*) [30–32]. The aleatoric or statistical uncertainty comprises the unknowns in the model that differ each time we run the same experiment and which cannot further be analyzed in terms of a model. Aleatoric uncertainty is relatively easily characterized depending on the application. It is often modeled in terms of a density function based on statistical evidence and propagated using the Monte Carlo method. More recently Karhunen–Loeve and polynomial chaos expansions have been used to model aleatoric uncertainty with a fair amount of success. Epistemic or systematic uncertainty, on the other hand, is a reducible type of uncertainty and can be further analyzed in terms of other upstream uncertainty sources. The analysis of epistemic uncertainty includes efforts to better understand the structure and processes of the system. The purpose of uncertainty quantification is to *reduce the epistemic uncertainty to aleatoric uncertainty* which in turn is simply modeled using probability density functions.

1.2.2 *Uncertainty Quantification*

Uncertainty quantification (UQ) aims at characterizing the uncertainty in a system within the framework of a model to understand the impact of uncertainty on system performance to reduce uncertainties and their impact on system performance. Uncertainty quantification can be taken up as an exercise of propagating the quantifiable (mainly aleatoric) uncertainties forward through the model to predict system performance based on uncertainties in inputs and parameters of the model in a process called forward uncertainty propagation. Alternatively, uncertainty quantification can perform an inverse assessment of model uncertainty and parameter uncertainty from experimentally observed variability in the system performance. Although inverse assessment has its value in iterating and refining the model, its potential can be better harnessed and it can better be integrated into a design flow in an iterative refinement process coupled with forward uncertainty propagation. Most of the recent UQ work in microfabrication technologies has explored forward uncertainty propagation techniques, which we also discuss later in this book.

1.2.2.1 Survey of Forward Uncertainty Propagation Methods

Uncertainty propagation (UP) process usually targets the evaluation of descriptive statistics of the outputs such as its lower-order moments including mean and variance. The statistical model of the output can then be used to evaluate output reliability, which is a useful system performance measure. In some applications, evaluation of a complete probability density function of the output is desirable.

The techniques of propagating uncertainty modeled as a probability distribution are largely classified into five categories [33]:

1. *Simulation methods* such as Monte Carlo (MC) simulations [34–36], importance sampling [37, 38], and adaptive sampling [39]
2. *Local expansion methods* such as Taylor series expansion and perturbation methods [34, 40]
3. *Functional expansion methods* such as Neumann expansion and orthogonal or Karhunen–Loeve expansions including polynomial chaos (PC) expansion and wavelet expansion [40, 41]
4. *Most probable point (MPP) methods* such as first-order reliability and second-order reliability methods
5. *Numerical integration methods* such as full numerical factorial integration and dimension reduction

The choice of a UP method depends on the characteristics of the problem at hand, such as the accuracy sought, the computational resources available, the cardinality of the random parameters, the probability distributions involved, and whether only various moments of the output are sought after or the complete distributions. Our method of choice, polynomial chaos expansion (PCE), is particularly helpful when a complete probability distribution of the output is desired. PCE outperforms other UP methods when there are fewer uncertainties but the degree of uncertainty is large [42]; however, its performance deteriorates when input distributions other than the normal distribution are considered [33].

Monte Carlo simulation, among other simulation methods, is the traditional technique applied to UP. MC methods are very popular because of the convenience they offer in being noninvasive and conceptually simple. Their implementation does not vary with the dimensionality of the problem and hence can be applied equally easily to models of different sizes and complexity [42]. MC simulation can be easily implemented on top of deterministic solvers, having identified and modeled the system uncertainties in terms of their probability density functions (PDFs), by running numerous forward simulations to collect output statistics. However, MC simulation methods require a large number of simulation runs to obtain accurate statistics and are therefore typically computationally expensive.

The polynomial chaos expansion (PCE) method has been based on Wiener's seminal work [43] where he used *Hermite polynomials* to model a stochastic system with Gaussian uncertainty. In the past 25 years, many researchers have used PCE as an effective computational tool in a variety of engineering applications, such as computational fluid dynamics [44, 45], structural reliability [34, 35], and microelectromechanical systems [46–48]. Ghanem and Spanos used Hermite polynomials to expand the finite element realization of stochastic differential equations in structural mechanics [40, 49]. Karniadakis and Xiu generalized the class of orthogonal polynomials used for PCE to propose *generalized polynomial chaos* (gPC) inspired by the probability density function (PDF) correspondence with the weight function Askey polynomials [41] which Xiu applied extensively to stochastic differential equations [41, 50, 51]. Solutions to polynomial expansion have been suggested using stochastic Galerkin [40], stochastic collocation [51], or stochastic testing [52–54].

1.2.3 Uncertainty Propagation in Microfabrication

As microfabrication technologies advance in crafting smaller features in fabricated devices, bringing the resolution down to a few nanometers, the fabrication process becomes more susceptible to process variation, leading to degraded device performance and potential failures. Uncertainty quantification (UQ) is particularly important for MEMS, where high micromachining tolerances reduce confidence in achieving design specifications. Whereas controlling the tolerances is not in the hands of the designer, modeling the uncertainties and considering them in the design process can raise confidence in achieving the desired specifications. Uncertainty in the context of the reliability of MEMS is related to either the lack of understanding of the underlying models (epistemic uncertainty) or the variability in fabrication processes and layout (aleatoric uncertainty). These uncertainties are among the biggest challenges to the reliability of MEMS. Uncertainty quantification aims at reducing the epistemic uncertainties to aleatoric uncertainties which in turn are represented as parameterized probability density functions. Industry is seeking new design automation tools and UQ strategies to model and simulate uncertainties and analyze their impact on device performance. Some typical examples of the different UQ engineering outcomes that the industry is interested in are [48]:

- **Yield or reliability analysis.** In the semiconductor industry *yield* refers to the percentage of chips in a finished wafer that passes all tests and is found to function properly as per their design specifications. The MEMS designers are interested in estimating the yield of the devices that they design. The yield estimate or the device reliability information is fed back into the design flow to help optimize designs for better yield and reliability.
- **Device performance tolerance and pricing.** Process variations impose uncertainties in device performance, which is interpreted as the expected tolerances of device performance. The classification grade and pricing decisions for the device depend on the performance statistics of the chip.
- **Hierarchical design.** Complex devices are often realized by putting together their individually designed components. In such situations, the yield and reliability of the device are determined by integrating the statistical information of each of the components.
- **Inverse problems.** The observed and measured variability in the performance of the fabricated device can help designers perform an *inverse assessment* of model uncertainty and parameter uncertainty and infer the distribution of some device parameters. This helps in improving system models for subsequent device designs.

References

1. Dimitrios Damianos, Jerome Mouly, and Pierre Delbos. *Status of the MEMS Industry - 2021 Report*. Tech. rep. Yole Developpement, 2021.
2. Wai-Kai Chen. *The VLSI handbook*. CRC press, 2016.
3. Anantha Chandrakasan et al. "Trends in low power digital signal processing". In: *Circuits and Systems, 1998. ISCAS'98. Proceedings of the 1998 IEEE International Symposium on*. Vol. 4. IEEE. 1998, pp. 604–607.
4. W Rhett Davis et al. "A design environment for high-throughput low-power dedicated signal processing systems". In: *IEEE Journal of Solid-State Circuits* 37.3 (2002), pp. 420–431.
5. KA Cook-Chennault, N Thambi, and AM Sastry. "Powering MEMS portable devices - a review of non-regenerative and regenerative power supply systems with special emphasis on piezoelectric energy harvesting systems". In: *Smart Materials and Structures* 17.4 (2008), p. 043001.
6. Shad Roundy. "On the effectiveness of vibration-based energy harvesting". In: *Journal of intelligent material systems and structures* 16.10 (2005), pp. 809–823.
7. Shad Roundy, Paul K Wright, and Jan Rabaey. "A study of low level vibrations as a power source for wireless sensor nodes". In: *Computer communications* 26.11 (2003), pp. 1131–1144.
8. Sang-Gook Kim, Shashank Priya, and Isaku Kanno. "Piezoelectric MEMS for energy harvesting". In: *MRS bulletin* 37.11 (2012), pp. 1039–1050.

9. Shashank Priya. "Advances in energy harvesting using low profile piezoelectric transducers". In: *Journal of electroceramics* 19.1 (2007), pp. 167–184.
10. Noël E Dutoit, Brian L Wardle, and Sang-Gook Kim. "Design considerations for MEMS-scale piezoelectric mechanical vibration energy harvesters". In: *Integrated Ferroelectrics* 71.1 (2005), pp. 121–160.
11. Dan Borza and Ioana Nistea. "Application of digital speckle interferometry for vibration analysis of a statically loaded vehicle tyre". In: *Journal of the Acoustical Society of America* 123.5 (2008), pp. 3868–3868.
12. Slobodan Stupar, A Simonovic, and M Jovanovic. "Measurement and analysis of vibrations on the helicopter structure in order to detect defects of operating elements". In: *Scientific Technical Review* 62.1 (2012), pp. 58–63.
13. Subhas Chandra Mukhopadhyay. "Wearable sensors for human activity monitoring: A review". In: *IEEE sensors journal* 15.3 (2015), pp. 1321–1330.
14. Long Yan et al. "A 0.5-μV_{rms} 12-μW Wirelessly Powered Patch-Type Healthcare Sensor for Wearable Body Sensor Network". In: *IEEE Journal of Solid-State Circuits* 45.11 (2010), pp. 2356–2365.
15. Jerald Yoo et al. "A 5.2 mW Self-Configured Wearable Body Sensor Network Controller and a 12μW Wirelessly Powered Sensor for a Continuous Health Monitoring System". In: *IEEE Journal of Solid-State Circuits* 45.1 (2010), pp. 178–188.
16. Jerome Mouly. *Wearables in Consumer and Medical Applications, 2020*. Tech. rep. Yole Developpement, 2020.
17. Renato Caliò et al. "Piezoelectric energy harvesting solutions". In: *Sensors* 14.3 (2014), pp. 4755–4790.
18. Shashank Priya and Daniel J Inman. *Energy harvesting technologies*. Vol. 21. Springer, 2009.
19. Jin Y Zhang et al. "Microstructure and piezoelectric properties of AlN thin films grown on stainless steel for the application of vibration energy harvesting". In: *Micro & Nano Letters* 7.12 (2012), pp. 1170–1172.
20. Asif Khan et al. "Piezoelectric thin films: an integrated review of transducers and energy harvesting". In: *Smart Materials and Structures* 25.5 (2016), p. 053002.
21. Mehmet Murat Altug Bicak. "Analytical investigation of squeeze film dampers". In: (2011).
22. TM Kamel et al. "Modeling and characterization of MEMS-based piezoelectric harvesting devices". In: *Journal of Micromechanics and Microengineering* 20.10 (2010), p. 105023.
23. CB Williams and Rob B Yates. "Analysis of a micro-electric generator for microsystems". In: *Sensors and Actuators A: Physical* 52.1-3 (1996), pp. 8–11.
24. Michael Goldfarb and Lowell D Jones. "On the efficiency of electric power generation with piezoelectric ceramic". In: *Journal of Dynamic Systems Measurement and Control* 121 (1999), pp. 566–571.
25. Henry A Sodano, Daniel J Inman, and Gyuhae Park. "A review of power harvesting from vibration using piezoelectric materials". In: *Shock and Vibration Digest* 36.3 (2004), pp. 197–206.
26. Steven R Anton and Henry A Sodano. "A review of power harvesting using piezoelectric materials (2003–2006)". In: *Smart materials and Structures* 16.3 (2007), R1.
27. S P Beeby, M J Tudor, and NM White. "Energy harvesting vibration sources for microsystems applications". In: *Measurement science and technology* 17.12 (2006), R175.
28. Rong Liu, Brad Paden, and Kimberly Turner. "MEMS resonators that are robust to process-induced feature width variations". In: *Journal of Microelectromechanical Systems* 11.5 (2002), pp. 505–511.
29. Marc C Kennedy and Anthony O'Hagan. "Bayesian calibration of computer models". In: *Journal of the Royal Statistical Society: Series B (Statistical Methodology)* 63.3 (2001), pp. 425–464.
30. Matthew Allen et al. "Reliability-based analysis and design optimization of electrostatically actuated MEMS". In: *Computers & structures* 82.13 (2004), pp. 1007–1020.
31. Armen Der Kiureghian and Ove Ditlevsen. "Aleatory or epistemic? Does it matter?" In: *Structural Safety* 31.2 (2009), pp. 105–112.

32. Hermann G Matthies. "Quantifying uncertainty: modern computational representation of probability and applications". In: *Extreme man-made and natural hazards in dynamics of structures*. Springer, 2007, pp. 105–135.
33. Jin Woo Lee et al. "Squeeze-film damping of flexible microcantilevers at low ambient pressures: theory and experiment". In: *Journal of Micromechanics and Microengineering* 19.10 (2009), p. 105029.
34. Ove Ditlevsen and Henrik O Madsen. *Structural reliability methods*. Vol. 178. Wiley New York, 1996.
35. Milik Tichy. *Applied methods of structural reliability*. Vol. 2. Springer Science & Business Media, 2012.
36. JE Hurtado and Alex H Barbat. "Monte Carlo techniques in computational stochastic mechanics". In: *Archives of Computational Methods in Engineering* 5.1 (1998), p. 3.
37. Rudiger Rackwitz and Bernd Flessler. "Structural reliability under combined random load sequences". In: *Computers & Structures* 9.5 (1978), pp. 489–494.
38. R.E. Melchers. "Importance sampling in structural systems". In: *Structural Safety* 6.1 (1989), pp. 3–10.
39. Christian G. Bucher. "Adaptive sampling - an iterative fast Monte Carlo procedure". In: *Structural Safety* 5.2 (1988), pp. 119–126.
40. Roger Ghanem and Pol Dimitrios Spanos. "Polynomial chaos in stochastic finite elements". In: *Journal of Applied Mechanics* 57.1 (1990), pp. 197–202.
41. Dongbin Xiu and George Em Karniadakis. "Modeling uncertainty in steady state diffusion problems via generalized polynomial chaos". In: *Computer methods in applied mechanics and engineering* 191.43 (2002), pp. 4927–4948.
42. Haiyan Cheng. "Uncertainty quantification and uncertainty reduction techniques for large-scale simulations". PhD thesis. Virginia Tech, 2009.
43. Norbert Wiener. "The homogeneous chaos". In: *American Journal of Mathematics* 60.4 (1938), pp. 897–936.
44. OM Knio and OP Le Maitre. "Uncertainty propagation in CFD using polynomial chaos decomposition". In: *Fluid Dynamics Research* 38.9 (2006), pp. 616–640.
45. Rafael A Perez. "Uncertainty analysis of computational fluid dynamics via polynomial chaos". PhD thesis. Virginia Tech, 2008.
46. Nitin Agarwal and NR Aluru. "Stochastic modeling of coupled electromechanical interaction for uncertainty quantification in electrostatically actuated MEMS". In: *Computer Methods in Applied Mechanics and Engineering* 197.43 (2008), pp. 3456–3471.
47. Xiaohui Guo et al. "Uncertainty quantification models for micro-scale squeeze-film damping". In: *International Journal for Numerical Methods in Engineering* 84.10 (2010), pp. 1257–1272.
48. Zheng Zhang. "Uncertainty Quantification for Integrated Circuits and Microelectromechanical Systems". PhD thesis. Citeseer, 2015.
49. Roger G Ghanem and Pol D Spanos. *Stochastic finite elements: a spectral approach*. Courier Corporation, 2003.
50. Dongbin Xiu, Ioannis G Kevrekidis, and Roger Ghanem. "An equation-free, multiscale approach to uncertainty quantification". In: *Computing in science & engineering* 7.3 (2005), pp. 16–23.
51. Dongbin Xiu. "Efficient collocational approach for parametric uncertainty analysis". In: *Commun. Comput. Phys* 2.2 (2007), pp. 293–309.
52. Zheng Zhang et al. "Stochastic testing method for transistor-level uncertainty quantification based on generalized polynomial chaos". In: *IEEE Transactions on Computer-Aided Design of Integrated Circuits and Systems* 32.10 (2013), pp. 1533–1545.
53. Zheng Zhang, Ibrahim Abe M Elfadel, and Luca Daniel. "Uncertainty quantification for integrated circuits: Stochastic spectral methods". In: *Proceedings of the International Conference on Computer-Aided Design*. IEEE Press. 2013, pp. 803–810.
54. Zheng Zhang et al. "Stochastic testing simulator for integrated circuits and MEMS: Hierarchical and sparse techniques". In: *Custom Integrated Circuits Conference (CICC), 2014 IEEE Proceedings of the*. IEEE. 2014, pp. 1–8.

Chapter 2
Review of Euler–Bernoulli Rectangular Beam Theory

Beams are among the most fundamental structural elements that are used in MEMS devices. Many complex MEMS structures can be modeled as beams connected in series or parallel, with or without a proofmass. Given that the boundary conditions at either end of a beam that occurs in MEMS devices are either free or clamped, the beams that commonly occur in a MEMS energy harvester are either a clamped-free cantilever beam or a doubly clamped beam. The modal analysis of the beam is performed as a differential eigenanalysis of the governing equation of the beam. The Euler–Bernoulli beam equation describes the spatial relationship between the beam's deflection and the applied load. It can be derived using Newtonian (force/moment balancing) or Hamiltonian (Lagrange) approaches [1]. In this chapter, we provide a review of the Euler–Bernoulli rectangular beam theory and use it to derive the modal response of the rectangular cantilever beam and the lumped model of its dynamic system.

2.1 Static Beam Equation

Beams have been used in construction and engineering since ancient times, and the interest in characterizing the behavior of beams was natural with the purpose of coming up with efficient and effective designs. Leonhard Euler and Daniel Bernoulli put together the theory to describe beam physics in 1750 which became known as the Euler–Bernoulli beam theory. The Euler–Bernoulli equation describes the relationship between the beam's deflection from its neutral position and the applied load:

$$\frac{d^2}{dx^2}\left(YI\frac{d^2w}{dx^2}\right) = q, \tag{2.1}$$

W. U. Syed, I. (Abe) M. Elfadel, *Tapered Beams in MEMS*,
https://doi.org/10.1007/978-3-031-66391-8_2

where the curve $w(x)$ is the deflection of the beam from its neutral position in the z direction at the position x in this one-dimensional model and q is a distributed load on the beam and may be a function of x, w, or other variables, Y is the elastic modulus, and I is the second moment of inertia of the beam's cross-section. The successive derivatives of w have important physical interpretations. The first derivative, dw/dx, is the slope of the beam, the second derivative is proportional to the bending moment, and the third derivative of this bending moment is the shear force in the beam. When YI is constant, which is often the case, the static beam equation can be simplified to

$$YI\frac{d^4w}{dx^4} = q.$$

A positive bending moment M produces compressive stress at the bottom surface and $dM = Qdx$ for moment equilibrium where Q is the shear force on the beam and $dQ = -qdx$ for force equilibrium when q is positive in the positive z direction. Integrating Eq. (2.1) we get

$$Q = -\frac{d}{dx}\left(YI\frac{d^2w}{dx^2}\right),$$

and integrating further we get

$$M = -YI\frac{d^2w}{dx^2}.$$

2.1.1 Deriving the Bending Equation

The static beam equation can be derived via force and moment balancing across a beam element under a small deflection assumption. Figure 2.1 shows how when a beam deflects, its deflection $w(x)$ is measured as the position of the point in x of the neutral axis of the beam relative to its position when the beam is in its neutral undeflected position.

Figure 2.2 shows a dx-sized element taken along the beam length with curvature $\frac{1}{\rho}$, subtending an angle $d\theta$ at the center of curvature. The length of the neutral axis is $\rho d\theta$. At a distance z below the neutral axis, the length of the element is $(\rho + z)\, d\theta$, and the strain on the beam element would then be

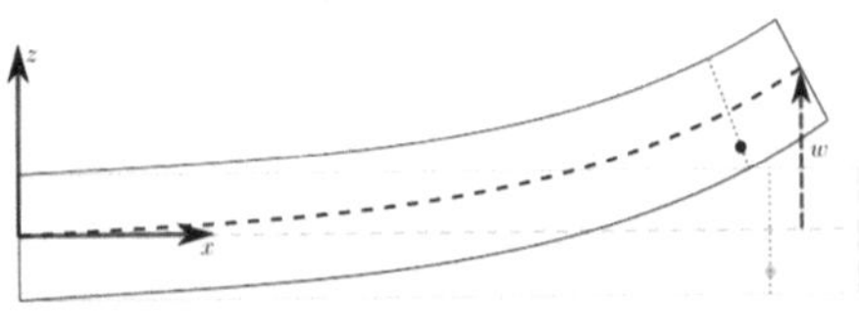

Fig. 2.1 Side view of beam deflection

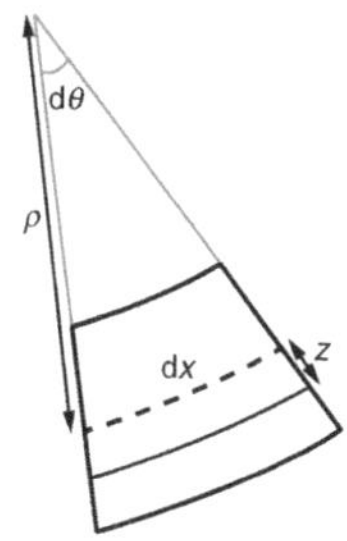

Fig. 2.2 A dx-sized element of deflected beam

$$\epsilon = \frac{(\rho + z - \rho)}{\rho d\theta} = \frac{z}{\rho}.$$

Hooke's law helps derive the stress on the element as $\sigma = Y\epsilon = Y\frac{z}{\rho}$ and the integrated force across the cross-sectional area element dA is

$$d\boldsymbol{F} = Y\frac{z}{\rho}dA\boldsymbol{e_x},$$

where $\boldsymbol{e_x}$ is the unit vector in the x direction. The differential bending moment would then be

$$d\boldsymbol{M} = -z\boldsymbol{e}_z \times d\boldsymbol{F} = -\boldsymbol{e}_y Y\frac{z^2}{\rho}dA. \tag{2.2}$$

Integrating Eq. (2.2) the bending moment is easily derived as

$$\boldsymbol{M} = \int d\boldsymbol{M} = -\boldsymbol{e}_y\frac{Y}{\rho}\int z^2 dA = -\boldsymbol{e}_y\frac{YI}{\rho}.$$

For small deflections where $\frac{dw}{dx}$ is small, the curvature is $\frac{1}{\rho} \approx \frac{d^2w}{dx^2}$, and

$$\implies \boldsymbol{M} = -YI\frac{d^2w}{dx^2}.$$

2.1.2 *The Dynamic Equation of a Beam*

Taking into account the inertial force generated by the bending motion of the beam, $\mu\frac{\partial^2 w}{\partial t^2}$, the balance of the force in Eq. (2.1) is generalized to

$$\frac{\partial^2}{\partial x^2}\left(YI\frac{\partial^2 w}{\partial x^2}\right) = -\mu\frac{\partial^2 w}{\partial t^2} + q, \tag{2.3}$$

where μ is the mass per unit length of the beam. This can also be derived using the Lagrangian method and Eq. (2.3) is also called the Euler–Lagrange equation.

2.2 Euler–Bernoulli Boundary Value Problem Formulation

The governing equation of motion for undamped free vibrations of a uniform beam is the dynamic Euler–Bernoulli beam equation or the Euler–Lagrange equation:

$$YI\frac{\partial^4 w_{rel}(x,t)}{\partial x^4} + m\frac{\partial^2 w_{rel}(x,t)}{\partial t^2} = 0, \tag{2.4}$$

where $w_{rel}(x,t)$ is the neutral axis displacement in the transverse direction, YI is the bending stiffness or flexural rigidity, and m is the mass per unit length of the beam. The clamped-free boundary conditions of the cantilever beam without any proofmass are defined at $x = 0$ and $x = l$. The geometric boundary conditions at $x = 0$ are given by BC_1 and BC_2 and the natural boundary conditions at $x = l$ are given by BC_3 and BC_4. These boundary conditions are given by

$$\begin{aligned}
&BC_1 : w_{rel}(0,t) = 0,\\
&BC_2 : \left.\frac{\partial w_{rel}(x,t)}{\partial x}\right|_{x=0} = 0,\\
&BC_3 : \left.\frac{\partial^2 w_{rel}(x,t)}{\partial x^2}\right|_{x=l} = 0, \text{ and}\\
&BC_4 : \left.\frac{\partial^3 w_{rel}(x,t)}{\partial x^3}\right|_{x=l} = 0.
\end{aligned}$$

2.3 Separation of Variables

$w_{rel}(x,t)$ is defined using the separation of variables as $w_{rel}(x,t) = \sum_{r=1}^{\infty}\eta_r(t)\phi_r(x)$. The rth mode is analyzed by substituting $w_{rel}(x,t) = \eta_r(t)\phi_r(x)$ in Eq. (2.4) to get

$$YI\eta_r(t)\frac{\partial^4\phi_r(x)}{\partial x^4} + m\phi_r(x)\frac{\partial^2\eta_r(t)}{\partial .t^2} = 0. \tag{2.5}$$

For a harmonic load at frequency ω_r, $\frac{\partial^2\eta_r(t)}{\partial t^2} = -\omega_r^2\eta_r(t)$, which when replaced above in Eq. (2.5) gives a differential equation in spatial domain describing the shape of the mode $\phi_r(x)$,

$$YI\frac{\partial^4 \phi_r(x)}{\partial x^4} - m\omega_r^2 \phi_r(x) = 0.$$

The substitution $\frac{m\omega^2}{IY} \rightarrow \left(\frac{\lambda_r}{l}\right)^4$ then obtains

$$\frac{\partial^4 \phi_r(x)}{\partial x^4} - \left(\frac{\lambda_r}{l}\right)^4 \phi_r(x) = 0.$$

This equation is solved for $\phi_r(x)$ using a trial solution $\phi_r(x) = c_1 \cos\left(\frac{\lambda_r x}{l}\right) + c_2 \cosh\left(\frac{\lambda_r x}{l}\right) + c_3 \sin\left(\frac{\lambda_r x}{l}\right) + c_4 \sinh\left(\frac{\lambda_r x}{l}\right)$. The boundary conditions BC_1 and BC_2 help eliminate c_3 and c_4 to get

$$\begin{aligned} \phi_r(x) =& c_1 \cos\left(\frac{\lambda_r x}{l}\right) - c_1 \cosh\left(\frac{\lambda_r x}{l}\right) \\ &+ (c_1 + 2c_2)\left(\sin\left(\frac{\lambda_r x}{l}\right) - \sinh\left(\frac{\lambda_r x}{l}\right)\right). \end{aligned} \tag{2.6}$$

2.4 Differential Eigenanalysis

Boundary conditions, BC_3 and BC_4, are applied to set up the system of equations for *differential eigenanalysis*,

$$c_1 \cos(\lambda_r) + c_1 \cosh(\lambda_r) + (c_1 + 2c_2)(\sin(\lambda_r) + \sinh(\lambda_r)) = 0, \text{ and}$$
$$c_1 \sinh(\lambda_r) + (c_1 + 2c_2)(\cos(\lambda_r) + \cosh(\lambda_r)) - c_1 \sin(\lambda_r) = 0,$$

which can be written in matrix form in terms of the two free parameters c_1 and c_2 as

$$\begin{bmatrix} \sin(\lambda_r) + \cos(\lambda_r) + \sinh(\lambda_r) + \cosh(\lambda_r) & 2(\sin(\lambda_r) + \sinh(\lambda_r)) \\ -\sin(\lambda_r) + \cos(\lambda_r) + \sinh(\lambda_r) + \cosh(\lambda_r) & 2(\cos(\lambda_r) + \cosh(\lambda_r)) \end{bmatrix} \begin{bmatrix} c_1 \\ c_2 \end{bmatrix} = 0.$$

To obtain nonzero values for the free parameters in this equation, the determinant of the system matrix must be zero. This condition helps establish the characteristic equation to evaluate the values of λ_r:

$$1 + \cos(\lambda_r)\cosh(\lambda_r) = 0.$$

The plot in Fig. 2.3 shows the zero crossings of the characteristic function that occur at each of the λ_r for increasing values of r. The characteristic equation must be solved numerically for its roots which are the first three wave numbers, $\lambda_1 =$

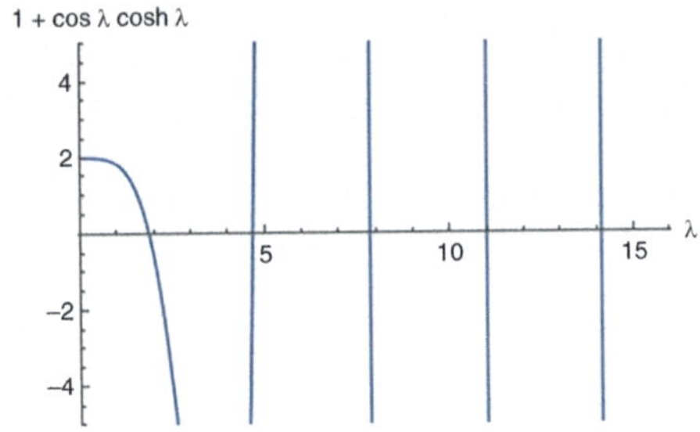

Fig. 2.3 Eigenvalues of the rectangular beam system

1.8751, $\lambda_2 = 4.6941$, and $\lambda_3 = 7.8548$. The eigenvalue of the system is computed using the following relationship:

$$\beta_{r,0} = \left(\frac{\lambda_r}{l}\right)^4.$$

The resonant frequency can then be evaluated using the relationship as follows:

$$\omega_r = \sqrt{\frac{YI_0}{m_0}\beta_{r,0}}.$$

2.5 Mass Normalization of the Mode Shapes

After applying BC_4 to the solution in Eq. (2.6), only one free parameter remains in the mode shape as

$$\phi_r(x) = c_1\left(\cos\left(\frac{\lambda_r x}{l}\right) - \cosh\left(\frac{\lambda_r x}{l}\right) + \sigma_r\left(\sin\left(\frac{\lambda_r x}{l}\right) - \sinh\left(\frac{\lambda_r x}{l}\right)\right)\right),$$

where $\sigma_r = \frac{\sin(\lambda_r)-\sinh(\lambda_r)}{\cos(\lambda_r)+\cosh(\lambda_r)}$. This mode shape is then mass-normalized by imposing the normalization constraint such that the mass inner product of the mode shapes is zero for different modes and unity when the mode numbers are the same, that is,

$$\int_0^l m\phi_i(x)\phi_j(x)\,dx = \delta_{ij},$$

where δ_{ij} is the Kroenecker delta. Mass normalization then results in the mode shape expression [2] as

$$\phi_r(x) = \frac{1}{\sqrt{lm}}\left(\cos\left(\frac{\lambda_r x}{l}\right) - \cosh\left(\frac{\lambda_r x}{l}\right) + \sigma_r\left(\sin\left(\frac{\lambda_r x}{l}\right) - \sinh\left(\frac{\lambda_r x}{l}\right)\right)\right).$$

2.6 Piezoelectric Energy Harvester Models

MEMS devices are described by an equation system that couples the different physics involved in the device. For energy harvesters, three physics domains are at play: structural, piezoelectric, and fluidic. The structural dynamics cause deflections and consequently stresses in the device structure, while the fluid damping resists the movement of the structure inside the device cavity. The piezoelectric effect translates the stress distribution to charge separation. There are damping phenomena other than fluidic damping including thermoelastic damping and anchor losses that have been explored in the course of this work. However, their contribution to the total damping of the device is insignificant at the cavity pressures of interest.

2.6.1 Lumped Parameter Models of MEMS Devices

Most MEMS devices exhibit the structural physics of a resonator where an effective mass m of the device is connected to its frame using a beam modeled as a spring with stiffness k. This mass vibrates at the modal frequency given by $\omega = \sqrt{\frac{k}{m}}$.

The lumped dynamic behavior of most MEMS devices that include inertial forces and viscous damping can, therefore, be approximately modeled as the spring–mass–dashpot system shown in Fig. 2.4 and can be described by the following equation:

$$m\frac{d^2x}{dt^2} + c\frac{dx}{dt} + kx = F, \tag{2.7}$$

which in state-space form is described as

$$\dot{x}_1 = x_2$$

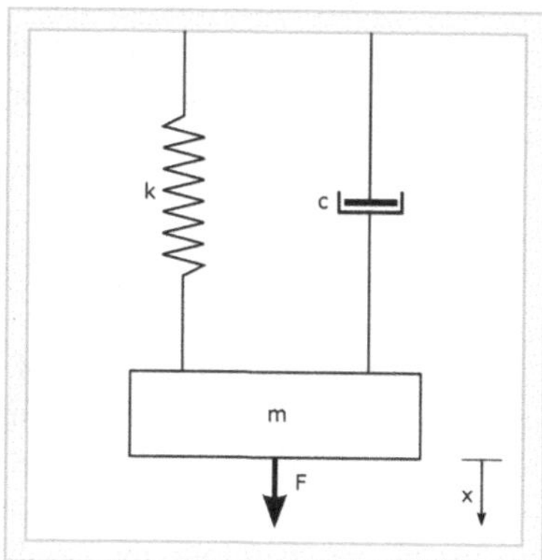

Fig. 2.4 Mass–spring–damper system

$$\dot{x}_2 = -\frac{k}{m}x_1 - \frac{c}{m}x_2 + \frac{1}{m}F.$$

For the state vector $\mathbf{x} = \begin{pmatrix} x_1 \\ x_2 \end{pmatrix}$, the state-space system is written as

$$\dot{\mathbf{x}} = \begin{pmatrix} 0 & 1 \\ -\frac{k}{m} & -\frac{c}{m} \end{pmatrix}\mathbf{x} + \begin{pmatrix} 0 \\ \frac{1}{m} \end{pmatrix}F. \tag{2.8}$$

Lumped models of specific MEMS devices are generated [3] by identifying the factors of mass m, damping c and stiffness k in Eq. (2.7) and the solution of the resulting dynamic system in Eq. (2.8) is sought.

2.6.2 *Piezoelectric Constitutive Equations*

The piezoelectric domain couples the mechanical and electrical behaviors of certain materials which cause charge separation in response to stresses in their lattice and vice versa. The electric polarization observed is proportional to the applied strain and is the result of the *direct* piezoelectric effect. The converse effect of mechanically actuating the material with an applied electric polarizing field is termed the *converse* piezoelectric effect. The linear constitutive equations for the unbounded piezoelectric continuum are given in tensor notation as [4–6]

$$S_{ij} = s^E_{ijkl}T_{kl} - d_{kij}E_k \tag{2.9}$$

$$D_i = d_{ikl}S_{kl} + \varepsilon^T_{ik}E_k, \tag{2.10}$$

where the subscripts i, j, k, and l take on the values $1, 2$, and 3. $\boldsymbol{T}$ is the stress tensor, $\boldsymbol{S}$ is the strain tensor, $\boldsymbol{E}$ is the electric field tensor, $\boldsymbol{D}$ is the electric displacement tensor, s^E is the elastic compliance matrix evaluated at the constant electric field, $\boldsymbol{d}$ is the matrix of piezoelectric strain coefficients, and ε^T_{ik} is the (i, k) entry in the permittivity matrix evaluated at constant stress. Equations (2.9) and (2.10) can be written in matrix form as

$$\begin{pmatrix} \boldsymbol{S} \\ \boldsymbol{D} \end{pmatrix} = \begin{pmatrix} s^E & \boldsymbol{d}^t \\ \boldsymbol{d} & \boldsymbol{\varepsilon}^T \end{pmatrix}\begin{pmatrix} \boldsymbol{T} \\ \boldsymbol{E} \end{pmatrix}.$$

A *unimorph* structure consists of one active layer, whereas a *bimorph* structure consists of two active layers. The unimorph energy harvester consists of a piezoelectric material sandwiched between two electrodes. A bimorph energy harvester consists of two unimorph layers that can be connected in different configurations. Unimorphs are, therefore, the building block structure of a piezoelectric transduction device.

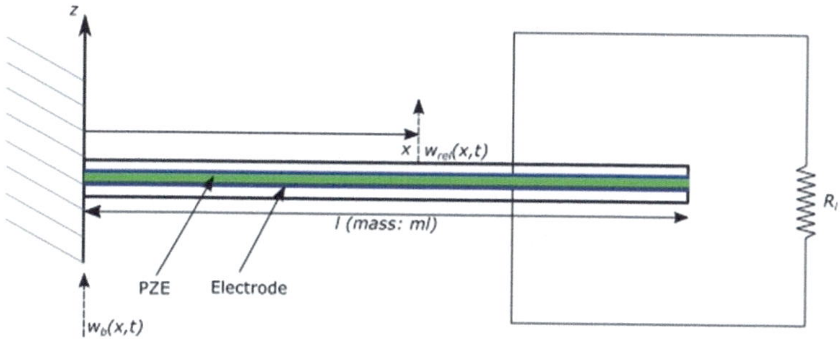

Fig. 2.5 Cantilever unimorph EH subjected to linear base excitation

The piezoelectric system is coupled to the structural deflection system through the axial deformation $S_1(, z, t)$ that is proportional to the curvature of the beam at position x. $S_1(, z, t)$ is given as

$$S_1(x, z, t) = -z\frac{\partial^2 w_{rel}(x, t)}{\partial x^2},$$

where w_{rel} is the deflection of the beam neutral axis in the z-axis.

2.6.3 *System Model of a Rectangular Beam PEH*

A cross-section of the side view of a unimorph cantilever beam is shown in Fig. 2.5. The density of the beam along the longitudinal direction is uniform. The analytical model of the EH is derived based on the piezoelectric constitutive equations given in Sect. 2.6.2 and the following assumptions [3]:

1. The cantilever beam behaves as an Euler–Bernoulli beam.
2. The external excitation from the air damping is small and negligible.
3. The damping forces derived from the mechanisms considered are all proportional to the bending stiffness and mass per length of the beam.
4. There is a uniform electric field across the piezoelectric layer.

2.6.3.1 Governing Equations

The structural governing equation of a cantilever beam with a rectangular profile based on the assumptions above is the dynamic Euler–Bernoulli beam equation with the damping term expressed as

$$\frac{\partial^2 M(x, t)}{\partial x^2} + c_a\frac{\partial w_{rel}(x, t)}{\partial t} + m\frac{\partial^2 w_{rel}(x, t)}{\partial t^2} = -m\frac{\partial^2 w_b(x, t)}{\partial t^2}, \quad (2.11)$$

where $M(x, t)$ is the internal bending moment at distance x from the anchor and time t, c_a is the viscous air damping coefficient, m is the mass per unit length of the beam, $w_b(x, t)$ is the base motion, and $w_{rel}(x, t) = w(x, t) - w_b(x, t)$ is the transverse beam deflection relative to its base. The internal moment including the damping and piezoelectric cross-coupling terms can be written as

$$M(x, t) = YI\frac{\partial^2 w_{rel}(x, t)}{\partial x^2} + c_s I\frac{\partial^3 w_{rel}(x, t)}{\partial x^2 \partial t} + \theta v(t),$$

where YI is the flexural rigidity of the beam, I is the moment of inertia, c_s is the strain rate damping coefficient, $v(t)$ is the voltage developed across the EH terminals, and θ is the electromechanical coupling coefficient. The mechanical system equation with piezoelectric coupling can then be obtained by substituting $M(x, t)$ in Eq. (2.11) to get

$$YI\frac{\partial^4 w_{rel}(x, t)}{\partial x^4} + c_s I\frac{\partial^5 w_{rel}(x, t)}{\partial x^4 \partial t} + c_a\frac{\partial w_{rel}(x, t)}{\partial t} + m\frac{\partial^2 w_{rel}(x, t)}{\partial t^2} + \theta v(t)\left(\frac{d\delta(x)}{dx} - \frac{d\delta(x-l)}{dx}\right) = -m\frac{d^2 w_b(x, t)}{dt^2}, \tag{2.12}$$

where $\delta(x)$ is the Dirac delta function which results from differentiating the Heaviside function introduced to define the extent of the electrodes. Integrating the second piezoelectric constitutive Eq. (2.10) over the area of the piezoelectric layer and differentiating with respect to t, the electromechanical equation can be derived as

$$\frac{v(t)}{R_l} + C_p^S\frac{dv(t)}{dt} + \int_0^l d_{31} h_{pc} b\frac{\partial^3 w_{rel}(x, t)}{\partial x^2 \partial t} dx = 0,$$

where C_p^S is the capacitance of the piezoelectric layer measured at constant strain and h_{pc} is the distance from the center of the piezoelectric layer to the neutral axis of the beam.

2.6.3.2 Mode Shapes

The simplified undamped and short-circuit analog of Eq. (2.12) then becomes

$$YI\frac{\partial^4 w_{rel}(x, t)}{\partial x^4} + m\frac{\partial^2 w_{rel}(x, t)}{\partial t^2} = -m\frac{\partial^2 w_b(x, t)}{\partial t^2}. \tag{2.13}$$

Employing the separation of variables where $w_{rel}(x, t) = \sum_{r=1}^{\infty} \eta_r(t)\phi_r(x)$, Eq. (2.13) can be solved for the mode shapes. The boundary conditions for the mode shapes of a cantilever beam as also suggested in Sect. 2.2 are given as [7, 8]

$$\begin{aligned} BC_1 &: \phi(0) = 0, \\ BC_2 &: \phi'(0) = 0, \\ BC_3 &: \phi''(l) = 0, \text{ and} \\ BC_4 &: \phi'''(l) = 0. \end{aligned} \tag{2.14}$$

Using the boundary conditions the mass-normalized solution for the mode shapes of the rectangular fixed-free beam derived by Erturk et al. [9] is given as

$$\begin{aligned} \phi_r(x) = &\frac{1}{\sqrt{lm}}\left(\cos\left(\frac{\lambda_r x}{l}\right) - \cosh\left(\frac{\lambda_r x}{l}\right)\right. \\ &\left.+\sigma_r\left(\sin\left(\frac{\lambda_r x}{l}\right) - \sinh\left(\frac{\lambda_r x}{l}\right)\right)\right), \end{aligned}$$

where l is the length of the beam, λ_r is the rth eigenvalue of the system, and $\sigma_r = \frac{\sin(\lambda_r)-\sinh(\lambda_r)}{\cos(\lambda_r)+\cosh(\lambda_r)}$. Also, note that $\eta_r(t)$ is the modal response function.

2.6.3.3 Electromechanical Coupled Modal Response ODE

Using the separation of variables as above the coupled electromechanical ordinary differential equations for the modal response of a rectangular cantilever beam EH can be given as [3, 9]

$$\ddot{\eta}_r(t) + 2\zeta_r\omega_r\dot{\eta}_r(t) + \omega_r^2\eta_r(t) + \chi_r v(t) = -f_r\ddot{g}(t) \tag{2.15}$$

$$\frac{v(t)}{R_l} + C_p\dot{v}(t) - \sum_{r=1}^{\infty}\chi_r\dot{\eta}_r(t) = 0, \tag{2.16}$$

where $g(t)$ is the linear motion of the base, ω_r is the resonant frequency of the structure, ζ_r is the damping ratio, χ_r is the coupling coefficient, and f_r is the forcing factor of mode r. χ_r and f_r can be written as

$$\chi_r = \int_0^l \theta\frac{d^2\phi_r(x)}{dx^2}dx, \text{ and}$$

$$f_r = \int_0^l m\phi_r(x)dx.$$

Notice that if the modes are well separated, then the summation of Eq. (2.16) will retain only the rth term. This system can be compiled as a state-space system with the state vector

$$\mathbf{q}(t) = \begin{pmatrix} \dot{\eta}(t) \\ \eta(t) \\ v(t) \end{pmatrix}.$$

The system for the rth mode can be written as

$$\boldsymbol{A}_r \dot{\mathbf{q}}(t) + \boldsymbol{B}_r \mathbf{q}(t) = \boldsymbol{F}_r,$$

where

$$\boldsymbol{A}_r = \begin{pmatrix} 1 & 2\zeta_r \omega_r & 0 \\ 0 & -\chi_r & C_p \end{pmatrix}, \text{ and}$$

$$\boldsymbol{B}_r = \begin{pmatrix} 0 & \omega_r^2 & \chi_r \\ 0 & 0 & \frac{1}{R_l} \end{pmatrix}, \text{ and}$$

$$\boldsymbol{F}_r = \begin{pmatrix} -f_r \ddot{g}(t) \\ 0 \end{pmatrix}.$$

This system can be solved for its transient solution with numerical solvers using the finite difference or finite element methods. Alternatively, an equivalent circuit model of this coupled electromechanical system can be constructed and solved using a circuit simulator.

2.6.3.4 Equivalent Circuit Model (ECM)

The equivalent circuit for a single mode of the rectangular beam energy harvester [3] is shown in Fig. 2.6. Note that the transformer models the coupling from the electrical to the mechanical domain. The left side of the transformer models the mechanical domain and the right side models the electrical domain. The piezoelectric layer capacitance is determined using the electrode area A_p, the piezoelectric layer thickness h_p, and the permittivity of the piezoelectric material measured at constant strain ε_{33}^S as

$$C_p = \frac{\varepsilon_{33}^S A_p}{h_p}.$$

The system Eqs. (2.15) and (2.16) help identify the equivalent circuit parameters R_r, L_r,C_r, and $V_r(t)$ shown in Fig. 2.6. Here R_r, L_r, and C_r are the equivalent

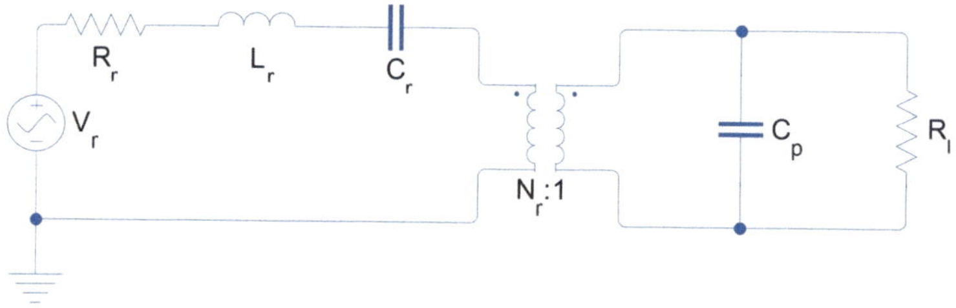

Fig. 2.6 Equivalent circuit of an energy harvester [3]

Table 2.1 Determining equivalent circuit parameters from electromechanical system model [3]

Equivalent circuit components at rth mode	Mechanical components
Charge: $q_r(t)$	Modal coordinate: $\eta_r(t)$
Current: $i_r(t)$	Modal velocity: $\dot{\eta}(t)$
Inductance: L_r	1 (mass normalized)
Resistance: R_r	$2\zeta_r\omega_r$
Capacitance: C_r	$\frac{1}{\omega_r^2}$
Voltage source: $V_r(t)$	Forcing function: $-f_r\ddot{g}(t)$
Ideal transformer ratio: N_r	Electromechanical coupling: χ_r

mechanical resistance, inductance, and capacitance of the rth mode, respectively, and $V_r(t)$ is the mechanical actuation expressed as its equivalent voltage. These parameters can be determined using the relationships with their mechanical counterparts shown in Table 2.1.

2.6.4 *Survey of Piezoelectric Energy Harvester Models*

A fixed-free *cantilever* energy harvester has been adopted as the structure of choice for energy harvesting research. In such a structure a cantilever is attached to the substrate structure at the base and is subjected to translational and rotational base excitation. Most researchers have considered harmonic translational base excitations for the sake of simplicity. When considering harmonic excitation, a single-degree-of-freedom (SDOF) model where only the tip deflection is considered allows the construction of a simple and tractable model.

Williams and Yates [10] captured some of the SDOF modeling and analysis principles of vibrational energy harvesters using an electromagnetic transducer for their evaluation. They concluded that the amount of power generated was proportional to the cube of the vibration frequency and that the device was likely to perform poorly at low frequencies. Du Toit et al. [11] applied Hamilton's principle [12] to the one-dimensional beam problem coupled with the piezoelectric physics to formulate an SDOF EH model. Du Toit [11], as well as Roundy [13], coupled the mechanical equilibrium and electrical loop equations through the piezoelectric constitutive relations to derive the lumped parameters representing the mechanical domain. However, they ignore the contribution of the distributed mass from the beam to the forcing amplitude in base excitation. This contribution is important for smaller proofmasses.

Sodano et al. [14] and Du Toit [15] employed the Rayleigh–Ritz discrete formulation originally derived by Hagood et al. [16] for piezoelectric actuation, to improve the EH model based on the Euler–Bernoulli beam theory. Consequent modeling attempts [17, 18] fell short of completing the resonance model with higher modes while also oversimplifying the electromechanical coupling force as viscous damping assuming that it is proportional to velocity. Erturk [8, 19] clarified some confusing modeling issues on piezoelectric energy harvesting and improved Du Toit and Sodano SDOF models by introducing correction factors to model outputs.

Erturk and Inman [9] developed a distributed parameter model for cantilever PEH to discover its exact analytical solution with the Euler–Bernoulli beam assumption. Liao and Sodano [20] developed a Rayleigh–Ritz model to predict power across a resistor considering a single excitation mode. Liao and Sodano [21] then extended their model to include a capacitor before the load in the conditioning circuit. Other researchers [22, 23] used nonlinear circuit techniques to develop a single-degree-of-freedom model for switched harvesting. Elvin and Elvin [24] proposed a general equivalent circuit model incorporating the backward coupling effect from multiple vibration modes and went on to develop a coupled finite element (FEA) and circuit simulation model to analyze piezoelectric energy harvesters. Yang and Tang [3] developed an accurate equivalent circuit model based on Erturk's model [9] and suggested a method to identify the equivalent circuit model (ECM) parameters from the theoretically formulated electromechanical EH system of simple structures as well as a procedure to discover these parameters from the finite element analysis results of complex structures. They further extended their equivalent circuit model to include multiple modes of vibration. Sahoo and Pandey [25] based their electromechanical model on the vibration model of Abrate [26] for nonuniform beams whose resonance frequency prediction error grows rapidly where taper factors exceed 0.5 [27].

References

1. William Weaver Jr, Stephen P Timoshenko, and Donovan Harold Young. *Vibration problems in engineering*. John Wiley & Sons, 1990.
2. TK Caughey and MEJ O'kelly. "Classical normal modes in damped linear dynamic systems". In: *Journal of Applied Mechanics* 32.3 (1965), pp. 583–588.
3. Yaowen Yang and Lihua Tang. "Equivalent circuit modeling of piezoelectric energy harvesters". In: *Journal of Intelligent Material Systems and Structures* (2009), pp. 2223–2235.
4. Shad Roundy. "On the effectiveness of vibration-based energy harvesting". In: *Journal of intelligent material systems and structures* 16.10 (2005), pp. 809–823.

5. Adrien Badel et al. "Single crystals and nonlinear process for outstanding vibration-powered electrical generators". In: *IEEE transactions on ultrasonics, ferroelectrics, and frequency control* 53.4 (2006), pp. 673–684.
6. AH Meitzler et al. *IEEE standard on piezoelectricity*. 1988.
7. H Thomas Banks and DJ Inman. "On damping mechanisms in beams". In: *Journal of Applied Mechanics* 58.3 (1991), pp. 716–723.
8. Alper Erturk. "Electromechanical modeling of piezoelectric energy harvesters". PhD thesis. Virginia Polytechnic Institute and State University, 2009.
9. Alper Erturk and Daniel J Inman. "A distributed parameter electromechanical model for cantilevered piezoelectric energy harvesters". In: *Journal of vibration and acoustics* 130.4 (2008), p. 041002.
10. CB Williams and Rob B Yates. "Analysis of a micro-electric generator for microsystems". In: *Sensors and Actuators A: Physical* 52.1-3 (1996), pp. 8–11.
11. Noël E Dutoit, Brian L Wardle, and Sang-Gook Kim. "Design considerations for MEMS-scale piezoelectric mechanical vibration energy harvesters". In: *Integrated Ferroelectrics* 71.1 (2005), pp. 121–160.
12. L. Meirovitch. *Fundamentals of Vibrations*. McGraw-Hill higher education. McGraw-Hill, 2001.
13. Shad Roundy and Paul K Wright. "A piezoelectric vibration based generator for wireless electronics". In: *Smart Materials and structures* 13.5 (2004), p. 1131.
14. Henry A Sodano, G Park, and DJ Inman. "Estimation of electric charge output for piezoelectric energy harvesting". In: *Strain* 40.2 (2004), pp. 49–58.
15. Noël Eduard Du Toit and Brian L Wardle. "Experimental verification of models for microfabricated piezoelectric vibration energy harvesters". In: *AIAA journal* 45.5 (2007), pp. 1126–1137.
16. Nesbitt W Hagood, Walter H Chung, and Andreas Von Flotow. "Modelling of piezoelectric actuator dynamics for active structural control". In: *Journal of intelligent material systems and structures* 1.3 (1990), pp. 327–354.
17. Shih-Nung Chen, Gou-Jen Wang, and Ming-Chun Chien. "Analytical modeling of piezoelectric vibration-induced micro power generator". In: *Mechatronics* 16.7 (2006), pp. 379–387.
18. JH Lin et al. "Modeling and simulation of piezoelectric MEMS energy harvesting device". In: *Integrated Ferroelectrics* 95.1 (2007), pp. 128–141.
19. Alper Erturk and Daniel J Inman. "On mechanical modeling of cantilevered piezoelectric vibration energy harvesters". In: *Journal of Intelligent Material Systems and Structures* (2008).
20. Yabin Liao and Henry A Sodano. "Model of a single mode energy harvester and properties for optimal power generation". In: *Smart Materials and Structures* 17.6 (2008), p. 065026.
21. Yabin Liao and Henry A Sodano. "Optimal parameters and power characteristics of piezoelectric energy harvesters with an RC circuit". In: *Smart Materials and Structures* 18.4 (2009), p. 045011.
22. Adrien Badel et al. "Efficiency enhancement of a piezoelectric energy harvesting device in pulsed operation by synchronous charge inversion". In: *Journal of Intelligent Material Systems and Structures* 16.10 (2005), pp. 889–901.
23. Elie Lefeuvre et al. "Piezoelectric energy harvesting device optimization by synchronous electric charge extraction". In: *Journal of Intelligent Material Systems and Structures* 16.10 (2005), pp. 865–876.
24. Niell G Elvin and Alex A Elvin. "A general equivalent circuit model for piezoelectric generators". In: *Journal of Intelligent Material Systems and Structures* 20.1 (2009), pp. 3–9.
25. Debesh Kumar Sahoo and Ashok Kumar Pandey. "Performance of non-uniform cantilever based piezoelectric energy harvester". In: *ISSS Journal of Micro and Smart Systems* (2018), pp. 1–13.
26. Serge Abrate. "Vibration of non-uniform rods and beams". In: *Journal of sound and vibration* 185.4 (1995), pp. 703–716.
27. Sajal Sagar Singh, Prem Pal, and Ashok Kumar Pandey. "Pull-in analysis of non-uniform microcantilever beams under large deflection". In: *Journal of Applied Physics* 118.20 (2015), p. 204303.

Chapter 3
Tapered-Beam Piezoelectric MEMS Energy Harvesters

Tapered piezoelectric energy harvesters take advantage of the tapered beam shape but in doing so the modeling complexity is greatly increased when compared to the nominal case of a rectangular piezoelectric energy harvester. This chapter presents the various benefits that can be gained from using a *tapered* beam shape both in terms of its structural merits and the charge separation capacity. Owing to the added complexity in modeling tapered beams, the design of tapered energy harvesters widely uses FEM simulations instead of formulating and solving a detailed mathematical model as is possible for rectangular energy harvesters shown in Sect. 2.6.3. To illustrate the traditional design methodology involving FEM simulations of the tapered device, the design, fabrication, and characterization steps of two generations of rectangular and tapered PEH devices are presented. The devices were fabricated at the foundry at the Institute of Microelectronics (IME), Singapore, and the design and characterization work was performed at the MEMS characterization facility at Masdar Institute (MI), Khalifa University.

3.1 Tapered Beams

A tapered cantilever (TC) is a commonly used structural element that has applications in diverse areas. When designing a TC to meet application requirements, the performance of the beam needs to be reliably and efficiently predicted which in most applications depends upon its modes of vibrations, their frequencies, and mode shapes. Modal analysis of tapered cantilevers is typically performed using the finite element method (FEM) which is inherently computationally intensive. In addition, the tapered edges complicate the meshing requirement thus adding further complexity to the analysis. When performing computer-aided design (CAD) of devices based on tapered cantilevers, we would like to be able to perform the modal analysis of the TC using an analytical model that has an analytical or a semi-

W. U. Syed, I. (Abe) M. Elfadel, *Tapered Beams in MEMS*,
https://doi.org/10.1007/978-3-031-66391-8_3

analytical solution. Such a solution will be fast to compute and should thus enable CAD scripts to search for optimal designs in a reasonable time. A faster solution of the device model further enables a much-needed uncertainty quantification for the design and yield analyses in a production environment.

3.1.1 Tapered Beams in Nature

To understand the value of tapered beams one can start by observing how natural selection for structural stability and other desired characteristics of overhanging structures in nature has resulted in the tapered shape. Researchers across various domains have been interested in developing models for tapered structures to understand the functionality of such features and work with them. Morgan et al. [1] used a series of untapered segments to analyze the vibrational behavior of tree trunks and branches subject to large deflections. The mammalian basilar membrane within the cochlea of the inner ear is tapered such that different parts of the cochlea are actuated at different auditory frequencies [2]. To gain insight into the mechanics of the inner ear, different researchers have modeled the structure using FEM [3]. Natural selection is believed to have resulted in the tapered whiskers of some terrestrial mammals presenting them with various functional advantages in addition to being structurally stable. The smaller tip helps with the sensing precision while keeping the resonant frequency relatively robust under tip breaks. Moreover, the maximum deflection angle for a tapered whisker is substantially smaller hence providing the whiskers with a higher spatial acuity [4].

3.1.2 Tapered Beams in Engineering

The structural reliability and better stress distribution along with other benefits make the tapered shape a popular choice in various overhanging components in building structures and machines. The tapered shape exploits the greater strength of a wider base and lower moments produced by the decreasing width going towards the tip resulting in a center of mass that is closer to the base. Kim [5] performed modal analysis of tapered traffic signal structures using both continuous and discretized models. The aircraft wings are tapered cantilever beams that have a cross section of an airfoil shape. Numerical methods such as FEM are used to solve for the wing bending often assuming it to be a tapered box beam [6, pp. 169–184]. Jet turbine engine blades are often twisted tapered beams. Bazoune et al. [7] explored the effect of tapering on such rotating beams. Ozgumus [8] and Huo [9] performed vibrational analyses on rotating doubly tapered Timoshenko beams.

3.1.3 Tapered Beams in MEMS

More recently, tapered cantilevers have been widely used as components of micro-electromechanical systems (MEMS). These are microscopic devices often with moving parts that are fabricated using modified semiconductor device fabrication technologies, normally used to make electronics. Tapered beams can be found in the inertial measurement units (IMUs) such as accelerometers, gyroscopes, and magnetometers where tapered edges of IMU fingers result in reduced squeezed-film damping. Tapered cantilevers are more relevant to the design of devices that employ beams as a part of their main operation. These include the atomic force microscope that uses a narrow, tapered microcantilever to sense and map out an object's surface [10]. Radio-frequency (RF) switches use capacitive or piezoelectric actuation to implement a mechanical switch for RF application. A switch with a tapered geometry is found to provide greater electrical isolation in its off-state [11] while also allowing for controlling the pull-in or actuation voltage of the switch by using the beam taper as a design parameter [12]. A direct motivation for the work presented in this book has been the modeling of a tapered cantilever-based vibration MEMS piezoelectric energy harvester (PEH). For a PEH, the tapered shape helps increase the maximum tolerable excitation amplitude of the device and consequently the maximum output power [13–15] while offering a higher structural reliability than a rectangular beam.

3.2 Tapered Beams in Piezoelectric Energy Harvesters

The design space within which piezoelectric energy harvesters are optimized includes the process stack, the piezoelectric layer configuration, the electrode segmenting and a polling strategy, a damping mitigation strategy, and the optimal shape and dimensions of the device within the available space. Since the process stack and the materials used by the foundry are not for the device designer to optimize, the most significant parameter that is tuned during the design process is the device's shape and size. Different shapes have been explored for a cantilever EH and researchers have pointed out some special benefits of using tapered shapes for cantilever energy harvesters.

3.2.1 Strain Distribution

The distribution of strain is an important factor in designing beam structures. In applications such as piezoelectric energy harvesting, the design goal is to maximize the net strain in the beam as it bends, or to achieve a specified tip deflection with minimal actuation thus requiring higher beam curvature. The piezoelectric constitutive equations (Eqs. 2.9 and 2.10) suggest that the total charge collected by the piezoelectric energy harvester increases with an increase in strain in the piezoelectric layer. Therefore, the task of the energy harvester designer is to maximize this strain to maximize power output. Cantilever structures fixed at one end and free at the other have been vastly recognized as beam structures that exhibit relatively high average strain on bending.

Whereas greater net stress in the TC structure is often sought by these applications, it must not be allowed to exceed the threshold of structural failure at any point along the beam. Fatigue-induced expected lifetimes are crucial figures of merit when designing devices for long-term use that operate close to resonance as an energy harvester would in order to maximize energy production. Structural integrity is often considered at the foundry where rules of thumb derived from experience are adhered to, such as the 500μ strain [16] limit to avoid brittle fractures in bulk materials. Structural robustness can be ensured by selecting the optimal shape that optimally distributes the bending stress over the beam span.

The traditional rectangular cantilever designs, unfortunately, exhibit large stress concentration at the fixed base, whereas the area around the free tip is hardly under any stress where the free end boundary conditions enforce zero curvature and consequently zero stress. Figure 3.1 shows the stress map of a rectangular cantilever with a tip mass generated from a FEM software, CoventorWare. It shows high stress around the clamped base for the rectangular cantilever and very low stress around the free tip.

Shape optimization of the beam structure will achieve a more uniform strain distribution and raise the average strain while reducing the maximum stress on the beam thus alleviating structural integrity challenges and increasing the structure's lifetime. One idea to reduce the stress is to balance the increasing moment the beam

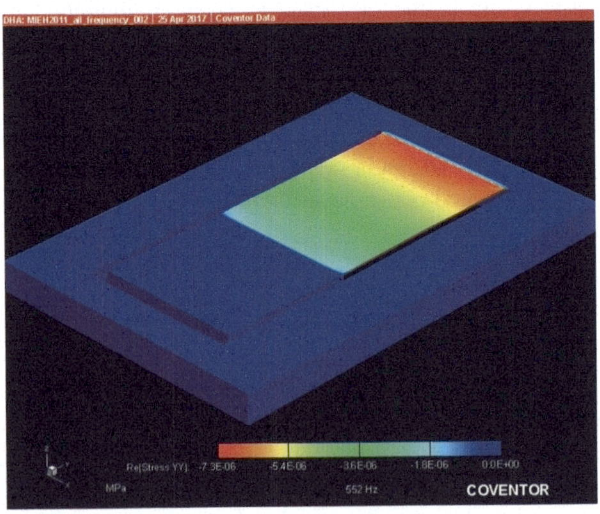

Fig. 3.1 Rectangular cantilever stress map close to resonance

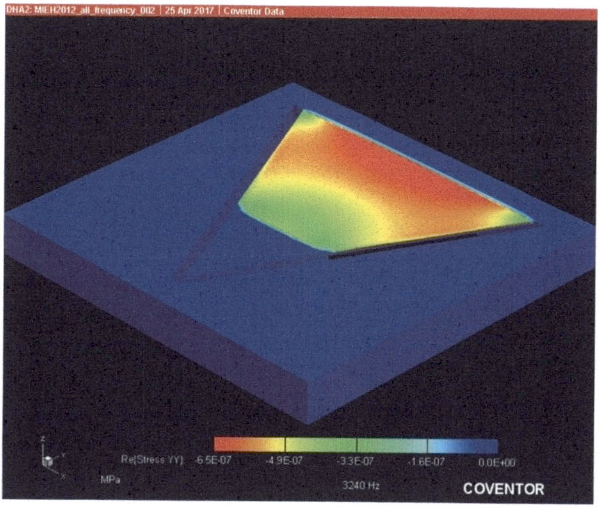

Fig. 3.2 Triangular cantilever stress map close to resonance

experiences going towards the base with an increasing width [14]. When the width increases linearly inwards towards the base, a *convergent* linearly tapered beam shape results. Figure 3.2 shows that the stress along the length of the triangular beam is better distributed than the rectangular beam.

Baker et al. measured and compared the strain distribution along the beam length for cantilevers at different tapers ranging from a rectangular cantilever to one that is almost triangular. For strain distributions along the beam length where the strain at the fixed end is the same for the different tapers, the results, shown in Fig. 3.3, indicate better-distributed strain as the taper of the beam increases.

Mateu et al. derived the average strain and maximum deflection expressions for straight and triangular cantilever beams [17]. Their final expressions are shown in Table 3.1 where h is the thickness of the beam, b_0 is the width of the rectangular cantilever and the width of the base of the triangular cantilever, l is the length of the beam, F is the point load, and Y is Young's modulus. The average strain expression shows that for a fixed maximum deflection as is the case when the device is capped, the rectangular beam has only 75% average strain as that of a triangular beam.

Goldschmidtboeing and Woias [13] also compared different beam shapes with respect to their energy harvesting efficiency which directly depends on the average strain in the beam and consequently on the average curvature. They define a relative mean curvature κ' to characterize the stress homogeneity:

$$\kappa' = \frac{|\kappa_{mean}|}{\kappa_{\max}}$$

where κ_{mean} refers to the mean curvature of the piezoelectric layer and $\kappa_{\max}$ is the maximum absolute value of the curvature. Best efficiency occurs when

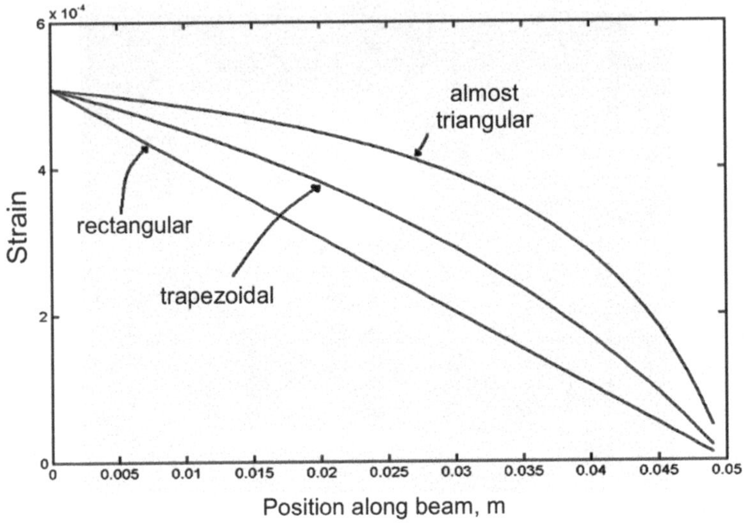

Fig. 3.3 Theoretical results: strain per length (at surface) [14]

Table 3.1 Average strain, S_x; deflection, $w(x)$; and maximum deflection, $w_{\max}$, for triangular cantilever and rectangular cantilever [17]

	Triangular cantilever	Rectangular cantilever
S_x	$\frac{3Fl}{Yb_0h^2} = \frac{w_{\max}h}{2l^2}$	$\frac{3Fl}{2Yb_0h^2} = \frac{3w_{\max}h}{8l^2}$
$w(x)$	$\frac{6Flx^2}{Yb_0h^3}$	$\frac{6Flx^2}{Yb_0h^3}\left(1-\frac{x}{3l}\right)$
$w_{\max}$	$\frac{6Fl^3}{Yb_0h^3}$	$\frac{4Fl^3}{Yb_0h^3}$

the mean curvature approaches the maximum curvature. They establish that the assertion of Matea et al. [17] that the triangular beam is optimally shaped for stress homogenization is particularly true for beams with large tip masses. Their results for normalized curvature as a function of normalized tip mass are shown in Fig. 3.4.

The argument that increased homogeneity of stresses in a triangular cantilever beam EH increases the harvester power is counterbalanced by the fact that the centroid of a triangular beam is placed at one-third the length of the beam from the fixed end as opposed to the halfway point in the rectangular beam EH. With a centroid closer to the fixed end, a triangular beam generates smaller moments under its own weight, and thus, the conversion of the mechanical excitation into electrical energy is compromised. A larger tip mass, however, can be used to compensate for the reduced moments and consequent loss in transduction efficiency. The net effect of the triangular shape is therefore small on the EH efficiency but its effect on the maximum tolerable excitation amplitude is large thus allowing for greater maximum output power [13]. Roundy et al. [16] also explored trapezoidal geometry for a bimorph beam EH and deduced that it can supply more than twice the energy as that obtained from a rectangular EH.

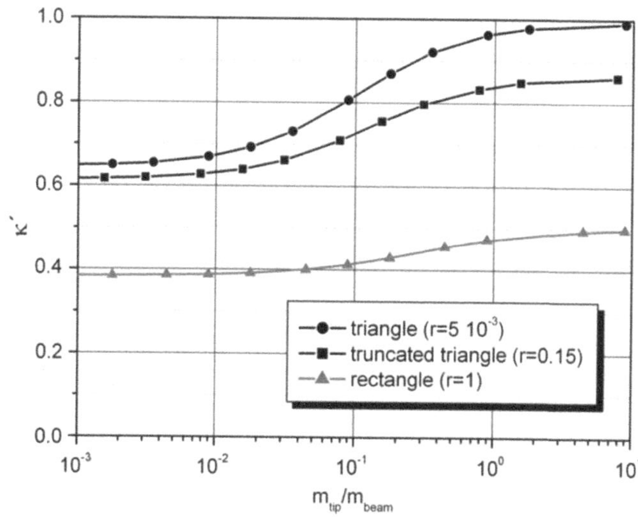

Fig. 3.4 Normalized curvature of the first resonant mode versus normalized tip mass for different beam shapes. r is the ratio between the widths at the free and fixed end [13]

3.3 Tapered Cantilever Model

An efficient vibration model and the modal solution of a TC are highly valuable for preparing a complete electromechanical model of the tapered PEH as they are to complete the operational model of other systems that employ the TC. Where analytical solutions are possible, they are preferred over numerical FEM solutions for their computational efficiency as well as their structure. Where devices and systems use a tapered component, the vibration physics of the tapered beam has to be coupled with other physics present in the system which gets easier with an analytical model. The structured closed-form analytical solutions also enable a deeper insight into the model dynamics that helps in developing more powerful analysis techniques.

The Euler–Bernoulli dynamic system model of the vibration problem of a TC and its modal equation can be set up by ignoring shear stress and assuming small deflection using the Newtonian (force/moment balancing) or Hamiltonian (Lagrange) approach [18] in a similar fashion as was done in Chap. 2 for the rectangular beam [19, 20]. The undamped governing equation of a tapered beam energy harvester under short circuit derived in Sect. 4.2 is

$$\begin{aligned} YI_0\left(1-\alpha\frac{x}{l}\right)\frac{\partial^4 w_{rel}(x,t)}{\partial x^4} &- \alpha\frac{2YI_0}{l}\frac{\partial^3 w_{rel}(x,t)}{\partial x^3} \\ + m_0\left(1-\alpha\frac{x}{l}\right)\frac{\partial^2 w_{rel}(x,t)}{\partial t^2} &= -m_0\left(1-\alpha\frac{x}{l}\right)\frac{d^2 g(t)}{dt^2}, \end{aligned}$$

where l is the length of the beam, YI_0 is the flexural rigidity of the beam at the fixed end, m_0 is the mass per unit length of the beam at the fixed end, and α is the base-width normalized tapering factor defined as $\alpha = \frac{b_0 - b_l}{b_0}$, where b_0 is the width of the beam at the anchor and b_l the width of the beam at the free end. The coupled governing equations and their solutions are derived in Chaps. 4 and 6.

3.3.1 *Survey of Tapered Cantilever Models*

Whereas the modal equation for a rectangular beam can be solved [21] in terms of trigonometric and hyperbolic trial functions, efficient modal solutions for the tapered beam structure are hard to derive as their direct solutions are expressed in terms of special functions [22, 23] such as the Meijer G and hypergeometric functions. The vibration of cantilever beams with constant thickness and linearly tapered width is often solved using FEM [24, 25] or other initial value numerical methods [26].

One can find solutions in the literature for beams that are tapered in different ways. The modal solution for torsional vibrations of a clamped-free conical rod is given [27, Sec. 3.1] in terms of Bessel functions of the first and second kind of order $\frac{1}{2}$ and $-\frac{1}{2}$. The bending vibration solution of a single-tapered beam with a taper in its height only is in terms of the first and second kind Bessel functions of order 1 [27, Sec. 3.4], whereas the bending displacement functions for double-tapered beams with identical taper ratio for the width and the height tapers are also based on Bessel functions [27, Sec. 3.5]. However, unlike the single-taper beams, these are order 2 Bessel functions. Attarnejad [28] presents the exact solution for the free vibration of a tapered beam with elastic end rotational restraints based on the Euler–Bernoulli theory. Note that the model is not for a prismatic beam with tapering width and the taper is applied in the thickness direction of the beam instead.

3.3.1.1 Approximate Solutions

In the absence of closed-form solutions, researchers have resorted to semi-analytical methods to evaluate approximate solutions of tapered beams. This included building TC models from parts consisting of rectangular beams. Nikolic and Salinic [29] approximated the tapered prismatic beam as multiple rectangular elastic beams arranged in steps. They further approximated each of the beam segments as three rigid beams connected through revolute and prismatic joints containing corresponding springs. They were able to approximate resonance frequencies of the beam which they validated against the numerical solution of [26]. However, the optimal positioning of the prismatic joints in their model remains an open question. Accounting for the shear stress using Timoshenko beam models, Cekus [30] approximated the tapered beam as segmented beams in a similar fashion. Abrate [31] transformed the motion equation of a tapered beam into that of a rectangular beam assuming that the mode-shape expression of a rectangular beam can be factored out of the tapered beam mode shape. The resonance frequency prediction error from Abrate's method grows rapidly where taper factors exceed 0.5 [32].

Semi-analytical methods such as the Adomian decomposition method, the differential transform method, and the homotopy perturbation method have been used to solve for the vibration dynamics of tapered beams, particularly where the nonlinear Timoshenko model is considered, albeit for loading conditions and taper that do not directly match our case. Mao and Pietrzko [33] performed free vibration analysis on a beam with exponential taper using Adomian decomposition. Ozgumus and Kaya [34] did the same for a rotating double-tapered Timoshenko beam featuring bending–torsion coupling using the differential transform method. Joubari et al. [35] used the modified iteration perturbation method to find the periodic solution of a cantilever with tapering thickness. Coskun et al. [36] compared solutions from the Adomian decomposition method, the variational iteration method, and the homotopy perturbation method to solve for the mode shapes and resonance frequencies of the uniform Euler–Bernoulli beam and a tapered cantilever rod. Bayat et al. [37] derived expressions for the resonance frequencies of a cantilever beam also with a linearly tapering thickness using homotopy perturbation on a nonlinear beam model.

Other semi-analytical methods aim to analytically solve the set of equations that are set up by applying numerical methods to the governing equation and the boundary conditions. These methods are fundamentally numerical and their solution quality depends on the sampling density or the mode shapes assumed. Bert and Malik [38] used the differential quadrature method to perform free vibration analysis of a beam with varying thickness along the beam length. Huo [9] uses the Galerkin method to discretize the rotating double-tapered cantilever Timoshenko beam model over the modal domain using the rectangular cantilever eigenfunctions of the Euler–Bernoulli beam.

Various techniques have been suggested to solve for the transverse vibrations of prismatic beams with varying widths assuming series solutions of different forms. Guo and Yang [39] proposed a mode-shape solution in the form of a power series in the system eigenvalue where the series coefficients are determined recursively and the eigenvalues evaluated using the boundary conditions on the mode-shape solution. Although their method claims to address all nonuniform beams, its only illustration was using the special case of a fourth-order taper [31] where the equation of motion transforms into that of a uniform beam. Hsu et al. [40] employed the Adomian modified decomposition method (AMDM) to develop a procedure to iteratively solve for the mode shapes and natural frequencies for the transverse vibrations of a beam that has a linearly tapering width and thickness. They, however, do not demonstrate the use of the method on beams with tapering width and constant thickness. On the other hand, Naguleswaran [41] used the Frobenius method to develop a power-series solution of the mode shape equation of the transverse vibrations of a Euler–Bernoulli beam with constant thickness and linearly varying width. The natural frequencies are computed numerically using an iterative process that starts with a trial value and proceeds with adjustments derived from the characteristic equation set up by the boundary conditions. More series terms were required for the higher modes and higher truncation factors. The author also reported that the frequency solutions were not satisfactory for truncation factors greater than 0.75 as the tapered beam approaches a rectangular beam.

Whereas most analytical and semi-analytical solutions available in the literature are those where the taper is transverse, i.e., in the thickness of the beam, the case for MEMS structures is that of a prismatic beam with constant height but tapering width for which an efficient, accurate, and robust analytical or semi-analytical solution that can work reasonably well for all taper values has been found lacking in the literature. Moreover, since we are only considering transverse beam actuation, the rotating beam models in the literature are not relevant to our case.

3.4 Finite Element Modeling and Simulation

MEMS+ and CoventorWare are FEA modeling and simulation tools that can be used to construct MEMS device models based on foundry-provided technology information including material properties and the process layer stack.

3.4.1 CoventorWare Modeling and Simulation

CoventorWare helps build a 3D model of the device using the material properties, process stack, and the device layout geometry ensuring that the modeling and simulation of the device are consistent with its design. Once meshed, various analyses can be performed on the 3D device model. Modal analysis helps compute the resonance frequencies and effective resonator mass for the various modes analyzed. The modal shapes are used in the mode-shape solvers to simulate squeezed-film damping (SFD) between the device and the capping layers [15]. Structural losses can also be estimated in the form of thermoelastic and anchor loss damping. Finally, the total estimated damping coefficients are fed into the harmonic analysis to estimate the deflection map under load conditions of interest.

3.4.1.1 3D Model Construction

Foundry data sheets help set up the materials file in CoventorWare and the process platform specifies the process steps used to construct the 3D model. The device layout file in .GDS or .CAT format can then be used alongside the materials and process files to construct the 3D model for analysis.

The device layout file is first preprocessed to simplify the structure to be simulated. Only the part of the device that requires simulation is cut out. At this stage, any corner rounding is removed to allow simpler meshing to be applied to the 3D model and 2D surfaces resulting in better mesh aspect ratios.

Any layers that do not take part in the physics pertaining to the simulations are removed. Once the 3D model is constructed, it is further simplified by modifying layer properties and merging layers to remove thin layers to further improve mesh aspect ratios. The contributions of the removed layers can be captured in the resulting structure using effective material properties for the merged material. For instance, effective elasticity modulus is computed as

$$E_{eff} = \frac{\sum_i E_i t_i^2}{\sum_i t_i^2},$$

where t_i is the thickness and E_i is the elasticity modulus of the ith layer. The effective density and Poisson ratio are computed as thickness-weighted averages of the property values of the individual layers. The simplified structure leads to a simplified mesh with better aspect ratios. The primary mesh-type limitation is imposed by the squeezed-film damping solver in CoventorWare which requires a hexahedral mesh on the surfaces that bound the squeezed film. While guaranteeing hexahedral mesh, the mesh size is carefully chosen to obtain reasonable mesh aspect ratios while managing a low number of elements for faster simulation.

The eventual mesh-type limitation is imposed by the SFD solver which requires a hexahedral mesh on the surfaces that constitute the squeezed film. The mesh type of choice for rectangular structures is the Manhattan mesh owing to its straightforward rectangular (hence hexahedral) elements that are used reliably by the different solvers. Where Manhattan mesh is not an option (e.g., triangular beams), an extruded mesh is utilized with preferably a "split-and-merge" algorithm which ensures that the mesh is hexahedral (Fig. 3.5).

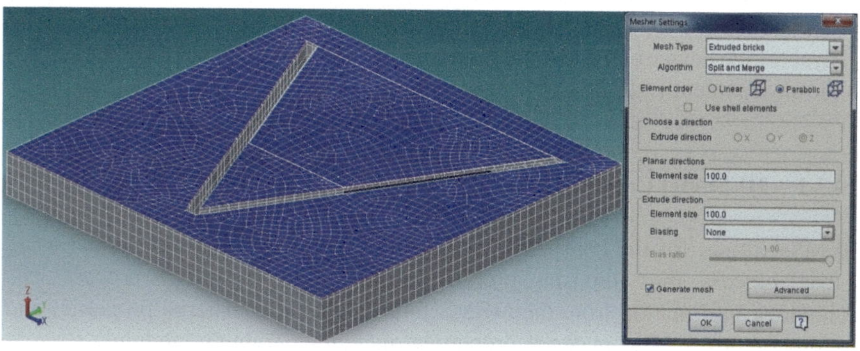

Fig. 3.5 Triangular EH device with extruded mesh

3.4.1.2 Simulation and Analyses

Modal analysis is performed to evaluate the resonance frequencies of the devices at different modes and the corresponding mode shapes along with their equivalent mass. To set up modal analysis in CoventorWare we set a fixed boundary condition on the bottom surface of the device walls. Squeezed and slide film damping analyses are set up by specifying the appropriate surfaces that take part in the damping physics. Squeeze film damping is the most important of the damping analyses at the pressure regimes of interest. The squeezed film surface boundary conditions are set to the sizes of the gaps between the device and the caps. Furthermore, CoventorWare computes the resistance at the edges of the squeeze film surfaces using a Reynolds equation-based model. Intuitively, the appropriate motion type for a cantilever would be the *modal shape*. However, for an EH design, where proofmass length is comparable to the beam length, we observe that the *rigid body* assumption gives results closer to the experimental value. We, therefore, chose to use *rigid body* assumption when computing squeezed-film damping from CoventorWare.

3.4.2 MEMS+ Modeling and Simulation

MEMS+ is a network-based MEMS modeling and simulation platform that simulates faster than the traditional field solver-based FEM software. The high simulation speed and parameterized models offered by MEMS+ allow MEMS designers to plug MEMS+ simulations into their design loop to optimize their designs for strategic goals. The models are constructed using parametric finite element components that comprise a network of connectors corresponding to the different physics domains including the mechanical, electrical, and fluidic domains that are being solved. MEMS+ is used to run modal and harmonic analysis on the model. Moreover, MEMS+ connects to MATLAB, Simulink, and Cadence, which can run a transient analysis of the device embedded in a larger system such as an electric circuit defined in the environment external to MEMS+.

3.4.2.1 3D Model Construction

MEMS+ allows for the construction of a parametric MEMS device model using a fixed process, thus imposing process constraints on the design process. The technology information from the MEMS foundry includes material properties and the process stack. The material properties, including Young's modulus, Poisson ratio, and density of the various layers, as well as the piezoelectric coefficients and permittivity of the piezoelectric layers, are entered in a MEMS+ materials (.MMDB) file. The process stack from the foundry is encoded into a process (.PROC) file. A parameterized device structure is then constructed either directly in MEMS+ or through a Matlab script. The design parameters representing device geometry such

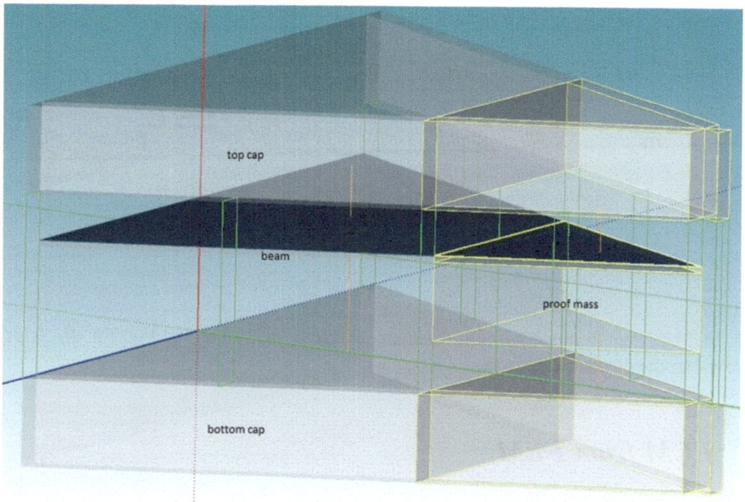

Fig. 3.6 MEMS+ model of tapered EH with a proofmass for SFD analysis. The green vertical lines are SFD connectors

as beam and proofmass lengths and widths, as well as taper angles, are left free to iterate over by the script. Boundary conditions are supplied to the mechanical model. Piezoelectric components are added along with electrode definitions. Beams are fragmented where multiple electrodes are required as in the case of doubly clamped beams. Finally, the squeezed film damping model is included depending on the cavities established by the device cross section as dictated by the process (Fig. 3.6).

3.4.3 Output Power Computation

The output power of the EH device is calculated for a load that matches the device input impedance by imposing current conservation (KCL) [42], $\frac{dq(t)}{dt} = \frac{V(t)}{R_l} + \frac{V(t)}{R_p} + C_p\frac{dV(t)}{dt}$, where $q(t)$ is the time profile of the collected charge obtained from MEMS+, $V(t)$ is the terminal voltage, R_p is the intrinsic resistance of the piezoelectric layer between the electrodes, C_p is the intrinsic capacitance of the piezoelectric layer between the electrodes, and R_l is the load across which the output power is computed as $P = \frac{V^2}{R_l}$. An optimal design solution is reached using the power calculation implemented in MATLAB and the device parameters are fine-tuned by exporting the full MEMS+ model into the Cadence Virtuoso environment for MEMS-IC co-verification and co-optimization.

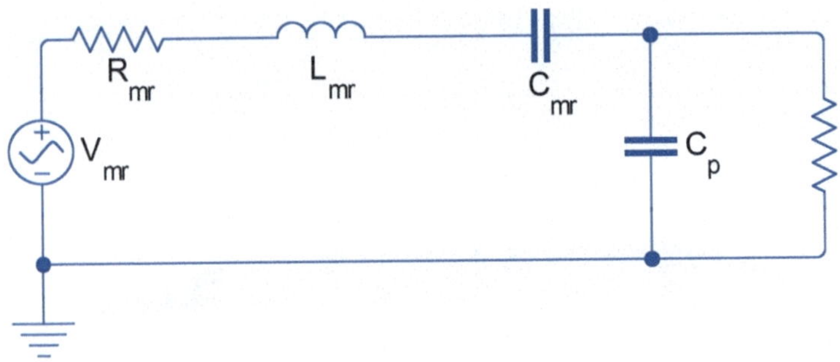

Fig. 3.7 ECM of the rth mode with the mechanical circuit parameters of Fig. 2.6, R_r, L_r, C_r, and $V_r(t)$ moved to the electric side [43]

3.4.4 ECM from FEM

For arbitrary EH geometry, it is difficult to construct analytical models and hence ECM derivation is not as straightforward. In such cases, the ECM parameters can be evaluated from an FEA (MEMS+) model of the device using a parameter identification process proposed by Yang et al. [43]. The parameters to be identified are the mechanical capacitance C_r, the mechanical inductance L_r, the mechanical resistance R_r, the coupling transformer turn ratio N_r, and the equivalent voltage $V_r(t)$ for the mechanical actuation (Fig. 3.7).

3.4.4.1 Identification of C_r, L_r, R_r, and N_r

Fixing the EH voltage to drive a current $-i(t)$ through the load and setting mechanical actuation to zero ($g(t) = 0$), we get the PEH admittance in the frequency domain as

$$Y = \frac{i}{V} = \frac{j\omega Q}{V} = j\omega\left(C_p + \sum_{r=1}^{\infty} \frac{\chi_r^2}{\omega_r^2 - \omega^2 + j2\zeta_r\omega_r\omega}\right).$$

Moving the components to the right of the ideal transformer with a turn ratio of N_r, we get

$$L_{mr} = \frac{L_r}{N_r^2},$$

$$R_{mr} = \frac{R_r}{N_r^2},$$

$$C_{mr} = C_r N_r^2, \text{ and}$$

$$V_{mr}(t) = N_r V_r(t). \tag{3.1}$$

Using Table 2.1 for the definitions of the circuit parameters and the relationships in Eq. (3.1), we get

$$Y = j\omega\left(C_p + \sum_{r=1}^{\infty} \frac{C_{mr}}{1 - C_{mr}L_{mr}\omega^2 + jR_{mr}C_{mr}\omega}\right).$$

For $\omega \rightarrow \omega_r$ the admittance can be shown to have two distinct components, Y_d and Y_{mot}, as

$$Y_d = j\omega C_{dr}, \text{ and}$$

$$Y_{mot} = \frac{j\omega C_{mr}}{1 - C_{mr}L_{mr}\omega^2 + jR_{mr}C_{mr}\omega},$$

where C_{dr} is the damped capacitance at the rth mode. The circuit parameters are identified from Y_{mot} [43] as

$$R_{mr} = \frac{1}{\max(Re(Y_{mot}))},$$

$$L_{mr} = \frac{R_{mr}}{\omega \min(Im(Y_{mot})) - \omega \max(Im(Y_{mot}))},$$

$$C_{mr} = \frac{1}{\omega_r^2 L_{mr}}, \text{ and}$$

$$N_r = \frac{1}{L_{mr}}.$$

Once the ECM is finalized using the circuit parameter values discovered from the FEM simulations, a larger circuit including the load is set up in a circuit simulator such as Cadence and simulated to perform power analysis. The revised design flow in Fig. 3.8 fits the ECM into the design loop along with the power computation by Cadence and Matlab.

3.5 Tapered Beam EH Design and Fabrication

Three generations of EH devices were designed through the course of this project using a process-aware design flow. The devices were designed in collaboration with IME and Global Foundries (GF), Singapore. Each generation was designed for an annual tape out at the IME foundry where the devices were fabricated based on

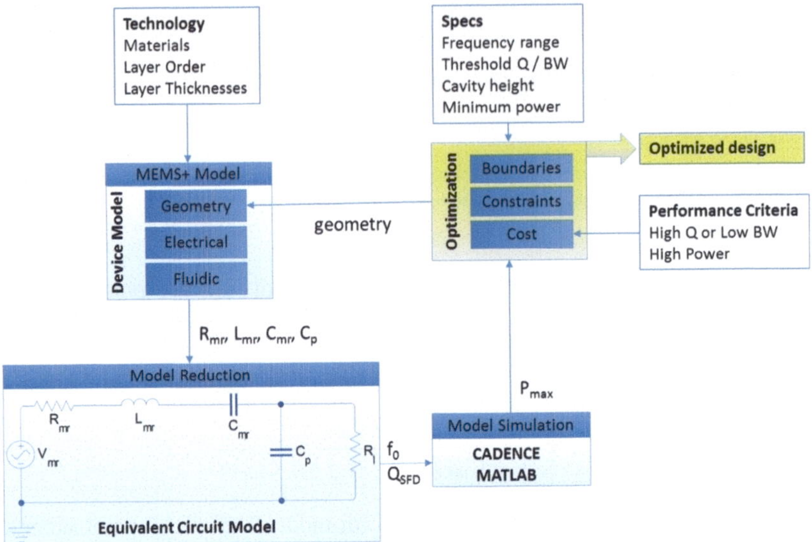

Fig. 3.8 Design optimization flow with ECM

IME's EH process platform. MEMS+, CoventorWare, and Cadence, as well as some Matlab code, described in Sect. 3.4, were used to validate the candidate designs.

3.5.1 Energy Harvester Process Platform

IME iterated the EH process platform for each EH generation with improvements geared towards structural robustness as well as greater efficiency of the EH device. The process is based on SOI wafer and uses AlN for the piezoelectric layer. A specialized under-etch creates a proofmass and, when combined with a top release, is capable of releasing the device comprising a beam and a proofmass. Figure 3.9 shows the nominal specifications for the different layers along with their foundry tolerances for the first-generation EH devices.

The main difference in the EH process platform for the second-generation devices (EV2) compared to the first-generation platform was an increased SOI device layer thickness of 30 μm and an increased cavity depth of 270 μm. The third-generation EH process platform increased the SOI device layer thickness to 50 μm and the cavity depth to 300 μm.

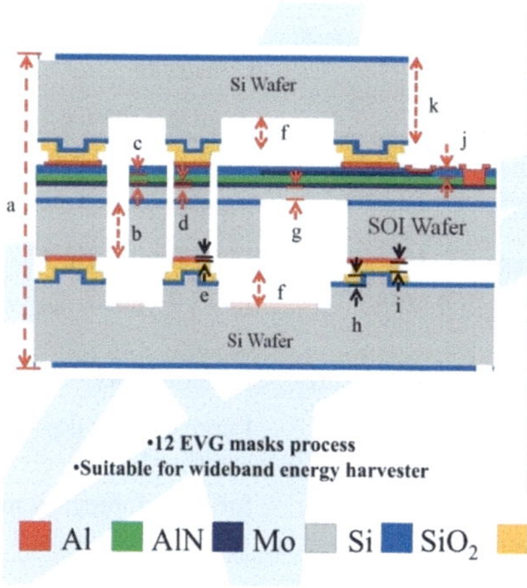

Dimension	Description	Unit	Specification		
			LL	Target	UL
a	Total device thickness	μm	1450	1500	1550
b	Proof mass thickness	μm	390	400	410
c	AlN thickness	μm	1.1	1.2	1.3
d	Mo thickness	μm	0.18	0.2	0.22
e	Al thickness	μm	0.63	0.7	0.77
f	Cavity depth	μm	90	100	110
g	SOI device layer	μm	19/14	20/15	21/16
h	Stand-off height	μm	2.3	2.5	2.7
i	Ge thickness	μm	0.35	0.4	0.45
j	SiO_2	μm	1.8	2	2.2
k	Cap wafer thickness	μm	540	550	560

Fig. 3.9 IME's first-generation EH process platform [44]

3.5.2 *Design Targets*

The primary design goal for the first-generation EH devices was to prepare diverse EH designs to test out the process platform's capabilities. Additionally, low-frequency operation with dominant resonant modes falling below 1 kHz was targeted. The design goals for the second-generation EH devices also included preparing diverse EH designs while repeating the rectangular and triangular cantilever EH designs from the first-generation devices to enable cross-release comparisons. Moreover, low-frequency operation close to 1 kHz or below was sought. A cavity depth of 270 μm constrained the maximum tip deflection before saturation occurs and a Q factor of 500 was targeted. Once the process platform was ready and stabilized, the third-generation EH devices were designed primarily to maximize the output power for the available die area. The design philosophy described below was generally followed in designing the optimal EH.

3.5.2.1 Design Philosophy

To fabricate an energy harvester that generates high power when excited with the expected vibrations in the environment, the device design process considers a large enough mass to allow complete device movement in the available space while keeping the spring stiffness and the device mass in line with the target primary mode resonance frequency. The transducer damping is kept small enough to allow the mass to move to the limits of its movement range which in our case are the inner surfaces of the upper and lower caps when actuated at typical vibration conditions. The design strategy described in Table 3.2 is generally followed when designing energy harvesters [45].

Table 3.2 EH device design philosophy

	Parameter	Goal	Design strategy	Constraint
1	Device mass, m_0	Maximize	$m_0 \uparrow \implies P \uparrow$	Available device volume
2	Tip deflection, $w_{\max}$	Maximize	Maximize cavity depth	Foundry rules
			$m_0 \downarrow \implies w_{\max} \downarrow$	Cavity depth
3	Resonance frequency, f_0	Target frequency (< 1 kHz)	$m_0 \uparrow \implies f_0 \downarrow$	Mass limited by #2
			$k \uparrow \implies f_0 \uparrow$	$k \propto \frac{1}{l_{beam}}$ and $l_{beam} < l_{device} - l_{proof\,mass}$
4	Damping, $c = c_s + c_a$	Minimize	Minimize cavity pressure	Better quality
			Minimize device area	Large enough device to satisfy #1–3
			Introduce slits and perforations	Structural integrity must not be compromised

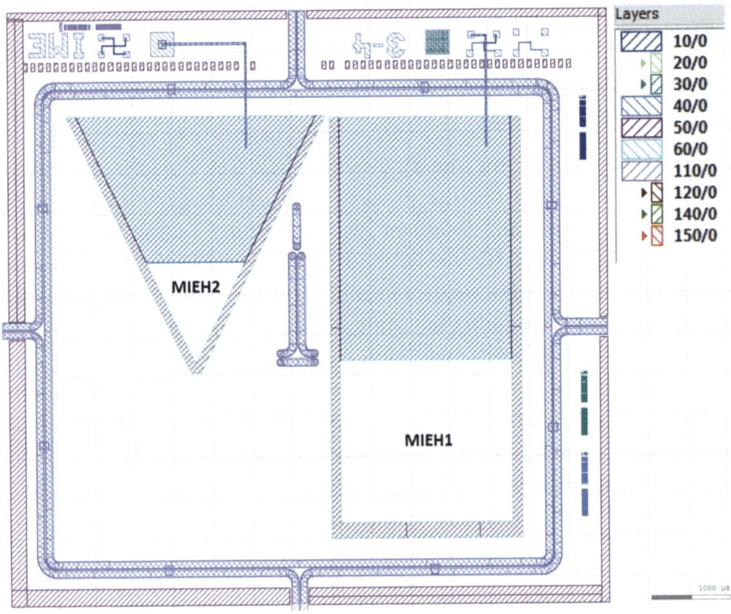

Fig. 3.10 Layout of first-generation devices

3.5.3 *Key Designs*

3.5.3.1 First-Generation Devices

Two of the first-generation device designs, the rectangular cantilever and the triangular cantilever energy harvesters, are of particular interest here. The layout of these devices is shown in Fig. 3.10.

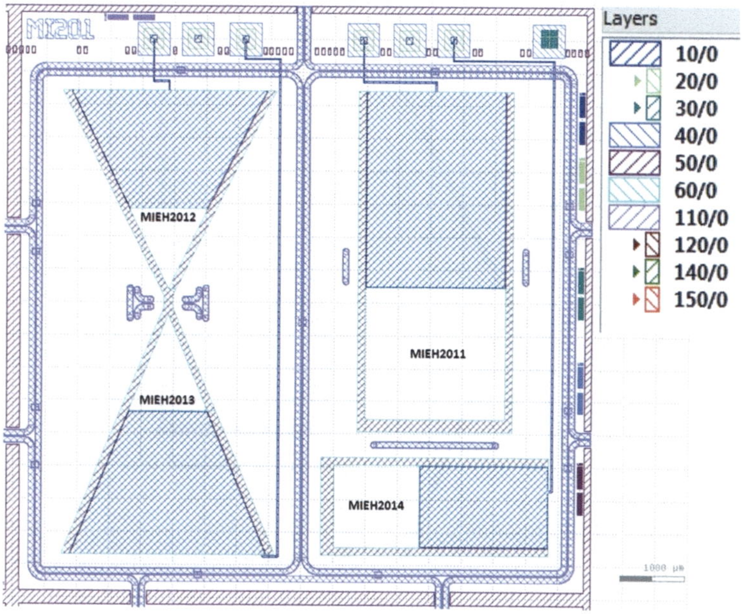

Fig. 3.11 Layout of some second-generation devices

Note that layer 60 is the bottom release and layer 50 is the top release, whereas layer 40 is the top electrode of the device. The device is released from its frame with trenches around it where layers 50 and 60 meet. Only layer 60 defines the beam part of the device with the remaining part being the proofmass.

3.5.3.2 Second-Generation Devices

The rectangular and triangular cantilever designs were repeated for the second-generation devices. These cantilever designs are shown in Fig. 3.11.

These devices were fabricated using the second-generation IME EH process platform. A digital microscope photograph of uncapped devices corresponding to the designs shown in Fig. 3.11 is shown in Fig. 3.12.

3.5.3.3 Third-Generation Devices

The third-generation devices were designed to maximize the output power generated for the given die area. After a comprehensive design search, our team settled on arrays of tapered beams that share one large proofmass. Since the scope of this book is isolated single-tapered cantilever beams with a limiting case of the rectangular beam, we do not present the third-generation devices here. It should be noted,

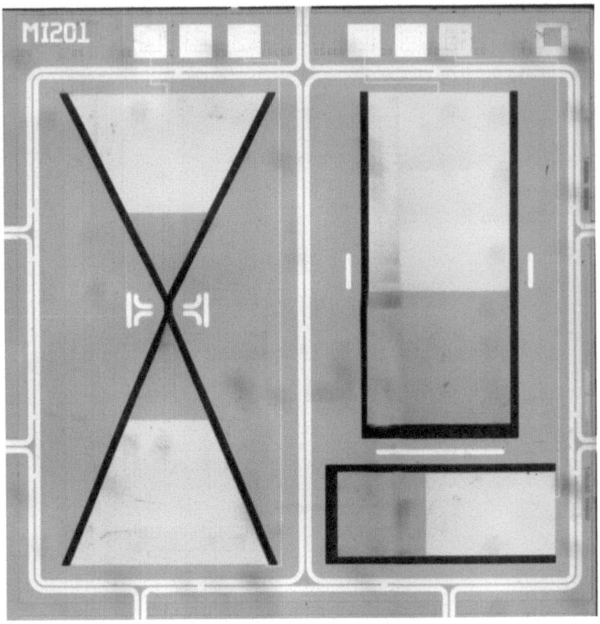

Fig. 3.12 A digital microscope photograph of the second-generation cantilever EH devices

however, that where the beam taper was taken up as a core topic in this book, the superiority of the tapered devices in terms of their transduction efficiency was also acknowledged in these designs.

3.6 Energy Harvester Characterization

3.6.1 Experimental Setup for Characterization

Figure 3.13 shows the EH characterization setup in the MEMS lab at MI. A function generator excites the shaker. The PCB with the device is installed on the shaker plate with an acrylic mount also designed and developed at MI. The EH output is connected via the PCB pads, wires, and BNC probe connectors to the acquisition system with B&K Shaker software for analysis. The EH signal conditioning circuit can also be interfaced with the EH PCB before analysis.

An acrylic device mount labeled A_2 in Fig. 3.14 was designed to place and hold multiple EH PCB's for simultaneous actuation by the shaker. The mount itself screws onto the top of the shaker reference accelerometer.

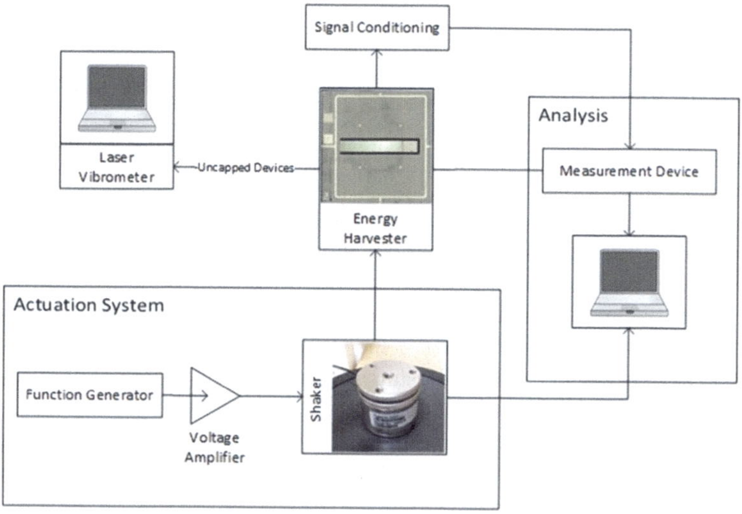

Fig. 3.13 EH characterization setup

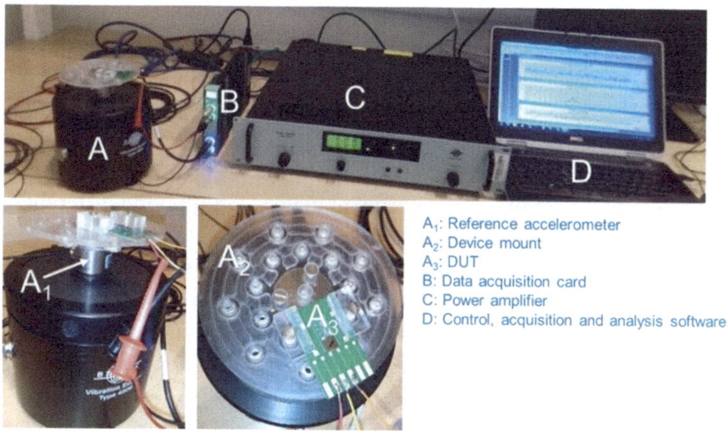

Fig. 3.14 Shaker setup for device characterization

3.6.2 Characterization Methodology

The objective of characterization is threefold. Firstly, visual inspection helps establish how well the fabricated devices match the device designs in their geometry and process stack. Secondly, visual inspection of the device deflection also helps discover the amount of residual stress in the stiffer layers during fabrication. Thirdly, the measured resonance frequency, Q-factors, open-circuit voltage, and output power are corroborated with the expected outputs derived from the device models.

3.6.3 Visual Inspection of the Device

The scanning electron microscope (SEM) helps inspect device cross section to compare the thicknesses of the various device layers with the nominal values of the process stack. The rigid frame of the device provides an easily accessible structure for such layer inspection.

Visual inspection of the devices using a digital optical microscope helps confirm device dimensions. Keyence microscope, an optical microscope equipped with a digital camera, was used to measure in-plane distances as well as depth between features in the view. Calibrated digital photographs of the uncapped devices taken from this microscope helped measure the dimensions of the device to compare them with nominal design values. Results of the device inspection with the optical Keyence microscope are presented in Sect. 3.7.1.

3.6.4 Residual Stress Analysis Methodology

To estimate the amount of residual stress that resides in the relatively stiffer layer, AlN in our case, we have to measure the deflection of the beam at its tip when at rest without actuation. The deflection thus measured is due to the warping experienced by the device when released with residual stress in its layers. *Keyance* digital microscope helps obtain a deflection measurement at the tip of the beam. The MEMS+ model of the same device can estimate the static beam tip deflection when a residual stress value is set up for the pertinent material. Supplying different residual stresses for the material and repeating harmonic analysis simulation in MEMS+ with zero actuation, one can discover the stress at which the tip deflection matches the measurement. The stress value found in this manner is an estimate of the residual stress of the layer. The analysis on a rectangular and a triangular beam is presented in Sect. 3.7.2.

3.6.5 Device Output Characterization Methodology

FEM models were first prepared as discussed in Sect. 3.4 to estimate the expected performance of the devices in terms of output voltage, power, and power density. During characterization, the device output was measured in the MEMS characterization lab at MI to validate the expected performance. The following measurements were used for EH characterization:

1. B&K shaker was used to sweep the mechanical actuation frequency at different acceleration levels and generate a spectrogram for open-circuit voltage. The resonance frequency and the Q-factors of the devices were inferred from these spectrograms.

2. The impedance analyzer and the LCR scope were used to measure the open-circuit impedance of the device. This impedance is used to provide an initial estimate of the matched impedance for optimal power output.
3. The optimal load and its corresponding power output were found by sweeping the load resistance across the EH terminals as the EH is actuated using the shaker.

3.7 Characterization Results of the Tapered Beam EH

3.7.1 Visual Inspection of the Device

SEM inspection was performed to confirm the layer thicknesses and explore possible defects. The walls and proofmass section of the uncapped devices were used to inspect the layers. The layer thicknesses were validated within a reasonable measurement error for devices of both generations. The inspection results for the first-generation devices are shown in Fig. 3.15.

3.7.2 Residual Stress Analysis

Residual stress analysis was performed as described in Sect. 3.6.4 for rectangular and triangular beam energy harvesters to estimate the residual stress in the AlN layer. A parameterized MEMS+ model was tuned for the residual stress that produced the Keyence microscope-measured beam deflection. The residual stress was validated using CoventorWare to produce the same beam deflection. Harmonic analysis was then run with zero load to obtain a deflection map that showed the structure warping owing to the residual stress configured in the model. The same analysis was repeated for the two generations of each of the devices.

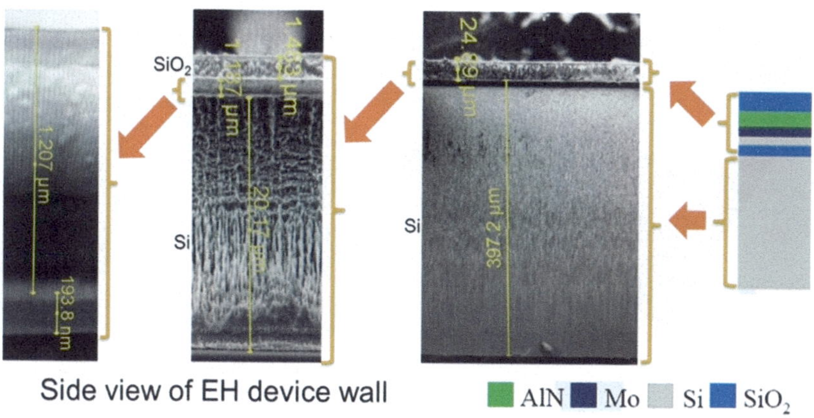

Fig. 3.15 SEM inspection of the EH layer stack

3.7.2.1 First-Generation Devices

For the straight cantilever shown in Fig. 3.16, the Keyence microscope measured the beam deflection at the proofmass tip as 164 μm. A tensile residual pressure of 256 MPa supplied to the AlN layer in the MEMS+ device model produced a maximum tip deflection of around 163.8 μm.

For the triangular cantilever shown in Fig. 3.17, the Keyence microscope measured the beam deflection at the proofmass tip as 68 μm. A tensile residual pressure of 114.5 MPa supplied to the AlN layer in the MEMS+ device model produced a maximum tip deflection of around 68 μm (Fig. 3.18).

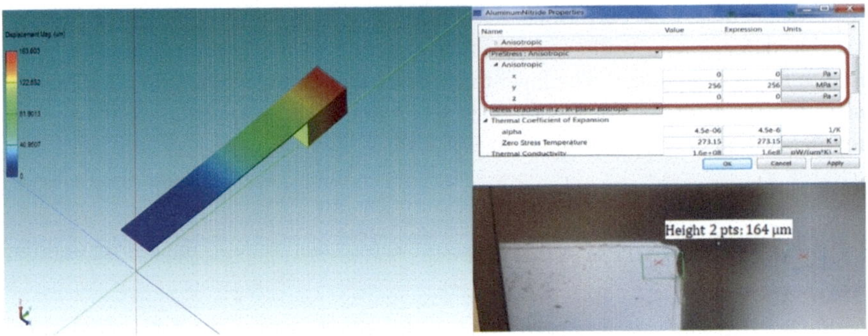

Fig. 3.16 Residual stress estimation via beam deflection analysis of first-generation rectangular cantilever

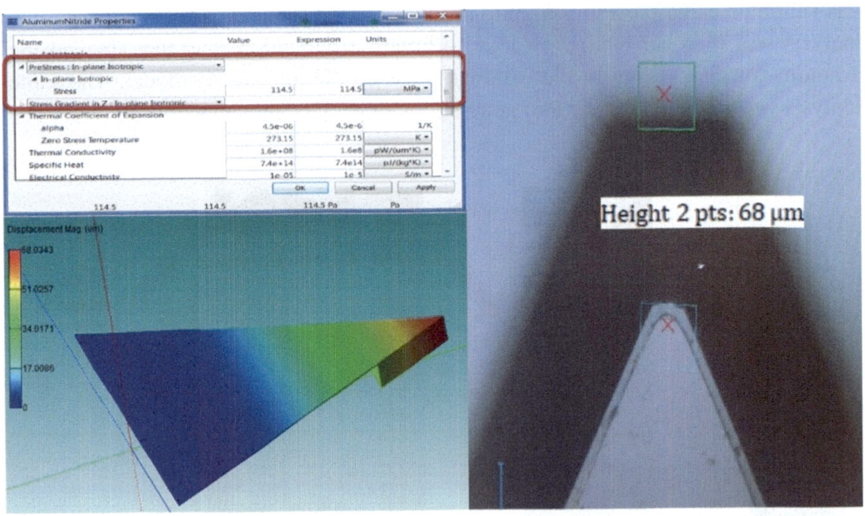

Fig. 3.17 Residual stress estimation via beam deflection analysis of first-generation triangular cantilever

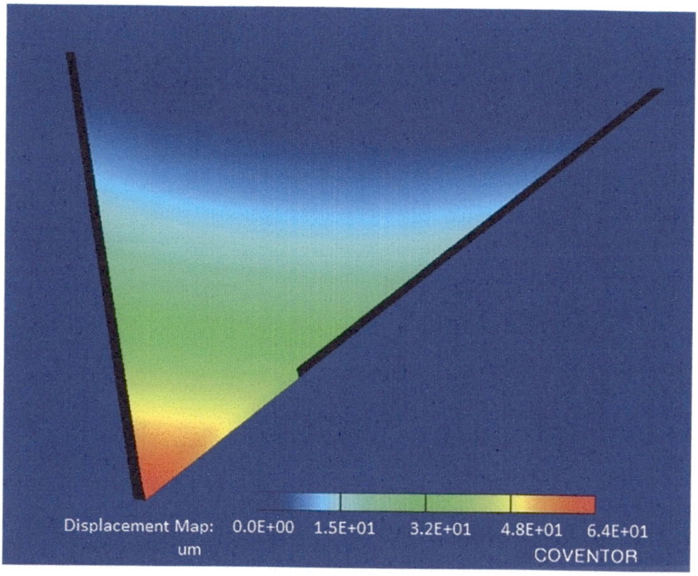

Fig. 3.18 CoventorWare estimated the deflection of the first-generation triangular cantilever beam under AlN residual stress of 114.5 MPa

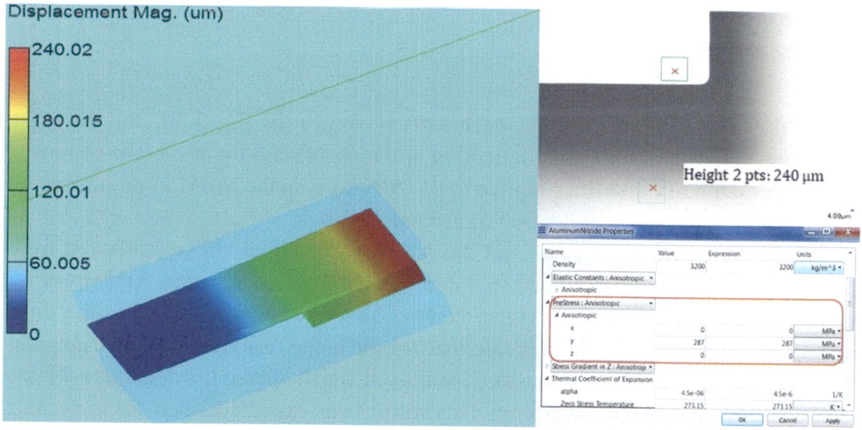

Fig. 3.19 Residual stress estimation via beam deflection analysis of second-generation rectangular cantilever

3.7.2.2 Second-Generation Devices

For the straight cantilever (MIEH2011) shown in Fig. 3.19, the Keyence microscope measured the beam deflection at the proofmass tip as 240 μm. A tensile residual pressure of 287 MPa supplied to the AlN layer in the MEMS+ device model produced a maximum tip deflection of around 240.02 μm.

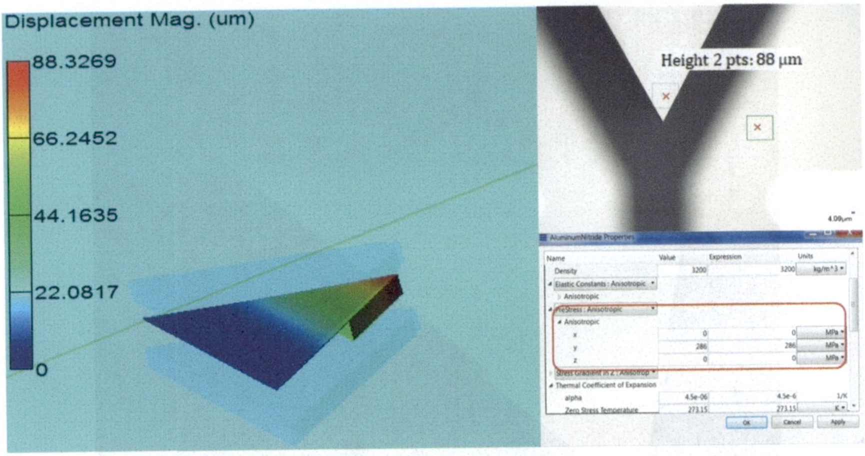

Fig. 3.20 Residual stress estimation via beam deflection analysis of second-generation triangular cantilever

Table 3.3 Measured tip deflections and derived residual stresses for rectangular and triangular devices

	Generation 1		Generation 2	
	Rectangular	Triangular	Rectangular	Triangular
Tip deflection (μm)	164	68	240	88
AlN residual stress (MPa)	256	114.5	287	286

For the triangular cantilever (MIEH2012) shown in Fig. 3.20, the Keyence microscope measured the beam deflection at the proofmass tip as 88 μm. A tensile residual pressure of 286 MPa supplied to the AlN layer in the MEMS+ device model produced a maximum tip deflection of around 88.3 μm.

The measured tip deflections and the AlN layer residual stresses inferred from these measurements are tabulated in Table 3.3. Though the reduced AlN stress for the first-generation triangular EH can be attributed to a modeling error at the time, the second-generation results indicate that for the same residual stress, a triangular cantilever of comparable dimensions results in much lower tip deflection of the beam and thus less device warping. This result is as expected since we know that the stresses are more evenly distributed in a triangular beam and hence a constant residual stress will cause a more even curvature of the beam resulting in less tip deflection.

3.7.3 Device Output Characterization

3.7.3.1 First-Generation Devices

We measured both the open-circuit voltage and the optimal power of the first-generation devices and the results are shown below. Figure 3.21 shows the measured

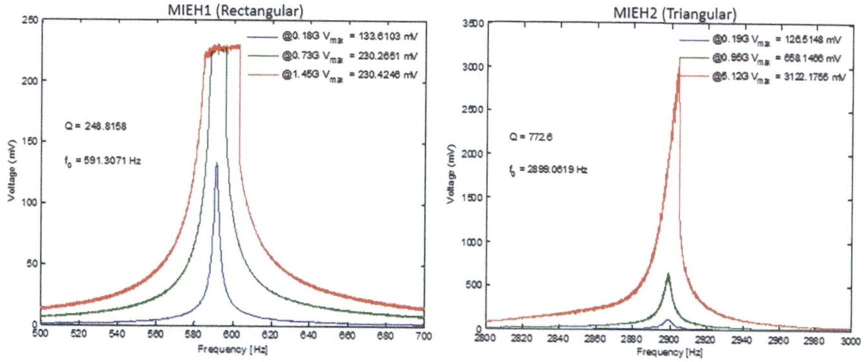

Fig. 3.21 Measured open-circuit voltage for the first-generation devices. MIEH1 is a rectangular device, and MIEH2 is a triangular device

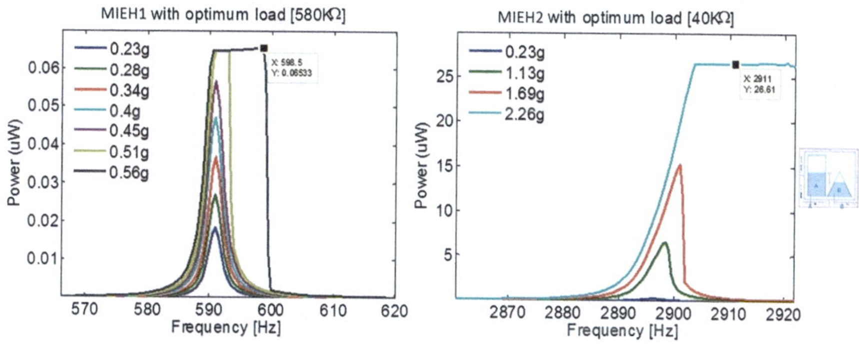

Fig. 3.22 Measured output power at optimum load for first-generation devices. MIEH1 is a rectangular device and MIEH2 is a triangular device

open-circuit voltage for the rectangular and triangular devices. Note that the open-circuit voltage at resonance for the triangular device is much higher than that for the rectangular device though the resonance frequency of the triangular device is also much higher. The design in Fig. 3.10 shows that the triangular device, MIEH2, has a smaller proofmass than the rectangular device, MIEH1, which explains the higher resonance frequency. The available space around the tapered beam provides an opportunity to increase the proofmass of the device to bring down the resonance frequency and bring up the power generated, a fact exploited in third-generation devices. Figure 3.21 indicates that the triangular EH can withstand higher acceleration before it saturates as opposed to the rectangular EH, enabling the triangular EH to reach a much higher saturation open-circuit voltage. Moreover, a lower area reduces damping and results in a higher Q factor for the triangular EH.

We swept the load resistance across the EH terminals to find the optimum load for maximum power output. In Fig. 3.22 we present the measured output power

Table 3.4 Measurement results for the first-generation cantilever devices

Device	Shape	Area (cm^2)	f_0 (Hz)	Q	V_{open} (V)	R_{opt} (kΩ)	P_{max} (μW)	$\frac{P_{max}}{A}$ (μW/cm^2)
MIEH1	Rectangular	0.1090	591	249	0.23	580	0.07 @0.5g	0.64
MIEH2	Triangular	0.0465	2899	773	3.12	40	26.75 @2g	575.27

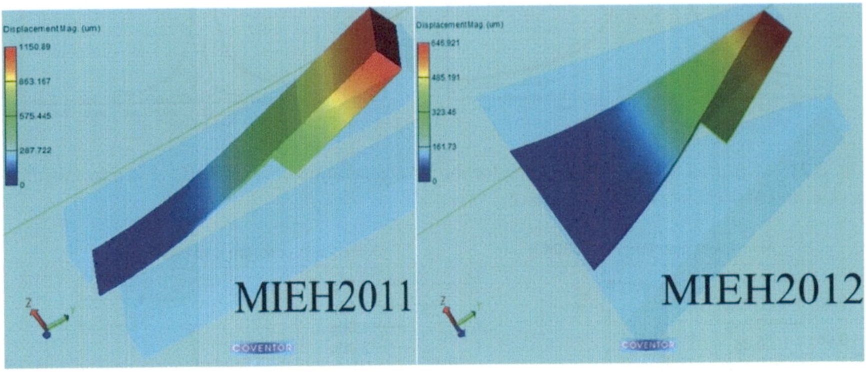

Fig. 3.23 Mode shapes from MEMS+ for MIEH2011 and MIEH2012

at this optimal load values for both the rectangular and the triangular structures for a range of actuation accelerations. Resistive optimal loading was considered due to its relative simplicity when compared to complex impedance loading which includes reactive impedance as well but will lead to a more accurate optimal power response. Note that the rectangular device MIEH1 saturates at around $0.5g$, whereas the triangular device saturates above $2g$.

The measurement results for the first-generation devices are summarized in Table 3.4. The area measured here is the device area including the beam and the proofmass. P_{max} is the maximum power generated by the device as it saturates when the oscillating beam structure starts hitting the top and bottom caps.

3.7.3.2 Second-Generation Devices

FEM Results

Modal analysis was performed both in CoventorWare and MEMS+. Modal analysis is used to look at the different vibration modes and the frequencies they occur at. A harmonic analysis performed with a wide frequency sweep allows us to locate the dominant mode. Modal analysis results from MEMS+ for a rectangular and a triangular cantilever device are shown in Fig. 3.23.

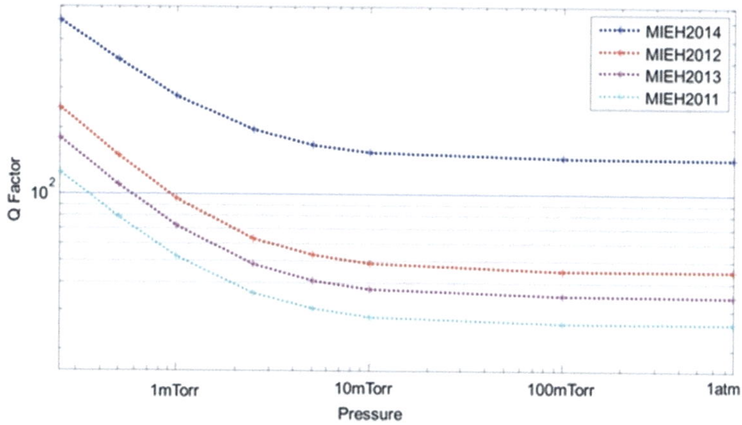

Fig. 3.24 Q *versus* P of the second-generation cantilever devices

Table 3.5 SFD results from CoventorWare for the cantilever devices

Device	Shape	f_0 (Hz) [CW]	f_0 (Hz) [MEMS+]	Q_{SFD} [CW]
MIEH2011	Rectangular	319	293	220
MIEH2012	Triangular	1884	1738	462
MIEH2013	Triangular	1317	1211	328
MIEH2014	Rectangular	709	651	981

SFD Analysis Results

We primarily focused on squeezed film damping (SFD) for the damping model since it is the dominant damping mode for higher-pressure regimes such as what we expect our devices to operate at. CoventorWare damping model was first cross-validated against other FEM tools such as Comsol and Ansys and later also validated against experimental data for a magnetometer comb oscillator. We could not, however, successfully validate the MEMS+ damping model. We, therefore, decided to use the CoventorWare SFD model to estimate the Q-factor at the operating pressure. The Q *versus* P SFD analysis obtained for the four devices using CoventorWare is shown in the log-log plot in Fig. 3.24.

The Q factors obtained from CoventorWare SFD models at 100 mTorr along with the modal analysis results of the second-generation cantilever devices are shown in Table 3.5.

Experimental Measurement Results

In order to characterize the pressure dependence of the SFD model for the EH device, we performed a pressure sweep to get a Q *versus* P plot. Overlaying the

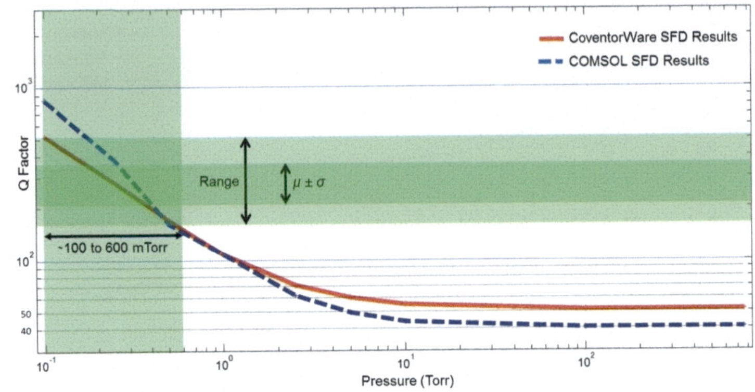

Fig. 3.25 Pressure regime inference from Q *versus* P analysis for the butterfly EH device

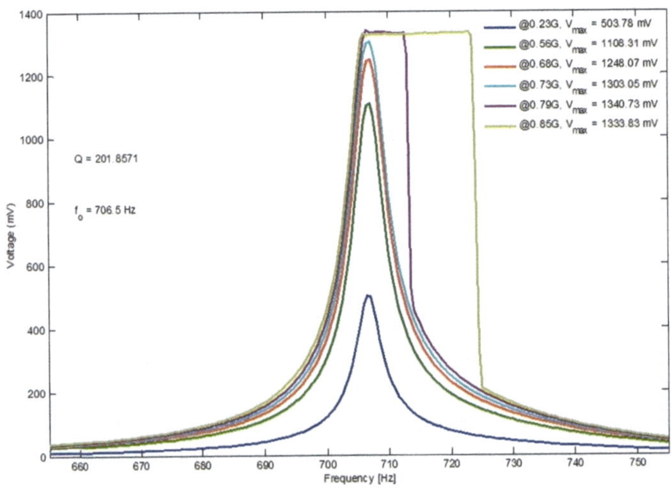

Fig. 3.26 MI2014 (rectangular) open-circuit voltage

statistical ranges of the tested Q factors of fabricated devices, we could infer the air pressure inside the cavity. The results shown in Fig. 3.25 obtained from CoventorWare as well as COMSOL indicate that the inferred cavity pressure ranges from 100 mTorr to 600 mTorr [46].

We measured both open-circuit voltage and the optimal power of the second-generation devices at a range of actuation accelerations. We used the load-sweeping process described earlier to find the optimal load for the second-generation devices The open-circuit voltage plots of the rectangular and triangular devices are shown in Figs. 3.26, 3.27, 3.28, 3.29, 3.30, 3.31.

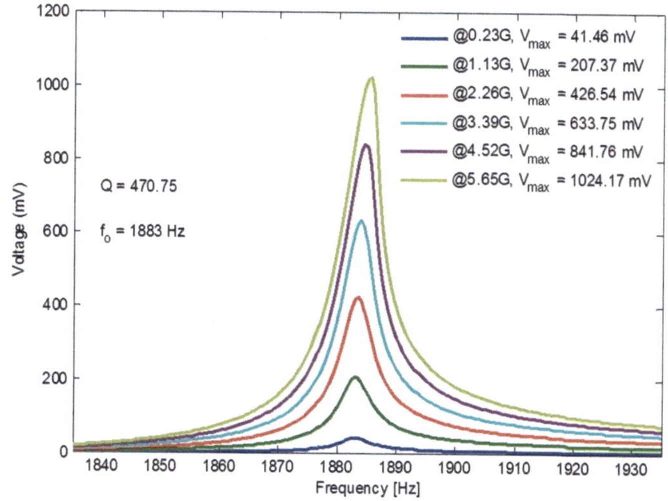

Fig. 3.27 MI2012 (triangular) open-circuit voltage

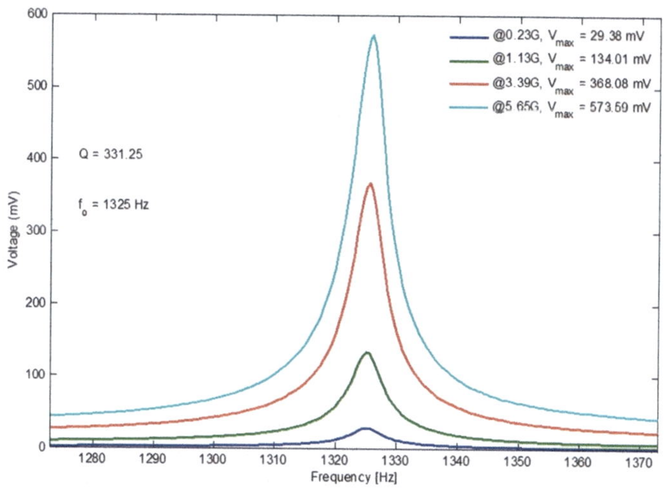

Fig. 3.28 MI2013 (triangular) open-circuit voltage

Figure 3.29 shows that the triangular devices show a nonlinear spring stiffening behavior as the resonance frequency increases with increased excitation acceleration. The power outputs measured at optimal resistive load for the triangular cantilever EH devices are shown in Figs. 3.30 and 3.31. Note again that the triangular EH does not saturate for an actuation acceleration of as high as $11.3g$. The rectangular EH, on the other hand, saturates for an acceleration of about $0.75g$. The measurement results for the second-generation devices are summarized in Table 3.6.

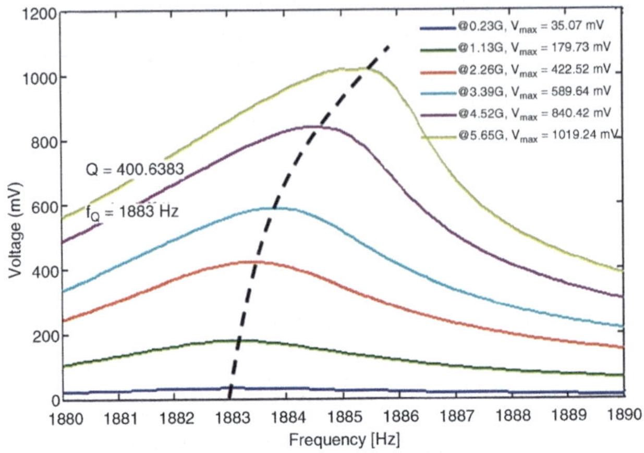

Fig. 3.29 MIEH2012 nonlinear response

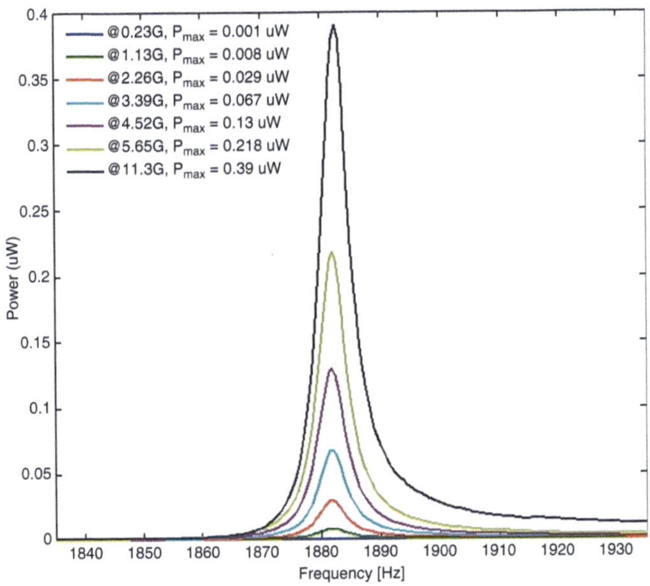

Fig. 3.30 MI2012 (triangular) output power across an optimum load of 100 kΩ

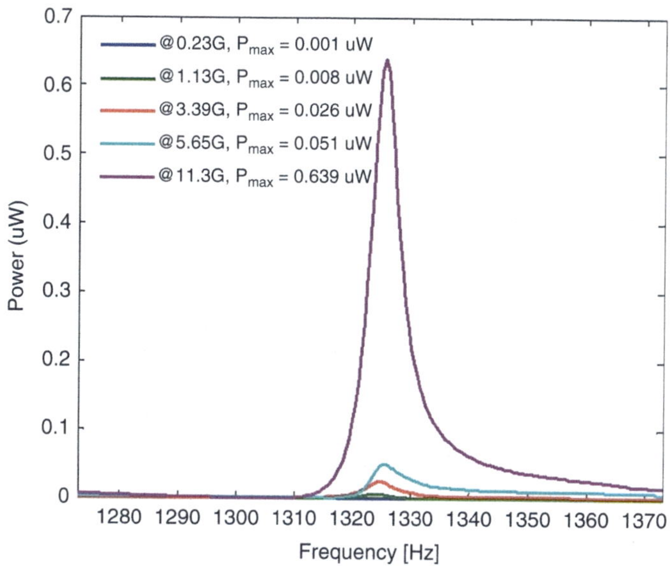

Fig. 3.31 MI2013 (triangular) output power across an optimum load of 280 kΩ

Table 3.6 Measurement results for the second-generation cantilever devices

Device	Shape	Area cm^2	f_0 Hz	Q	V_{open} V	R_{opt} kΩ	$P_{\max}$ μW	$\frac{P_{\max}}{A}$ μWcm^{-2}
MIEH2012	Triangular	0.046	1883	470.75	1.02	100	0.39	8.39
MIEH2013	Triangular	0.056	1325	331.25	0.57	280	0.64	11.45
MIEH2014	Rectangular	0.042	706.5	201.86	1.33	380	1.03	24.52

3.8 Conclusion

The characterization results are used to establish some of the benefits of tapered beams. The residual stress analysis discussed in Sect. 3.7.2 shows that the triangular beam exhibits less warping than the rectangular beam at about the same stress levels. Less static tip deflection will allow a greater range of motion for the tapered device to capture a greater dynamic range for excitation acceleration. This fact is also evident from the first-generation EH device power analysis results where the output from the triangular beam EH is shown to reach higher power levels. Characterization results also indicate that the triangular cantilever energy harvesters have a nonlinear response for higher values of excitation acceleration where their resonance frequency increases with increasing excitation acceleration. The assumption of small bending angles when deriving the bending moment for most analytical models in the literature is the reason why this nonlinearity is not incorporated in these models. This nonlinearity is expected to be a significant challenge for model validation.

References

1. J Morgan and MGR Cannell. "Structural analysis of tree trunks and branches: tapered cantilever beams subject to large deflections under complex loading". In: *Tree Physiology* 3.4 (1987), pp. 365–374.
2. Luis Robles and Mario A Ruggero. "Mechanics of the mammalian cochlea". In: *Physiological reviews* 81.3 (2001), pp. 1305–1352.
3. Namkeun Kim, Kenji Homma, and Sunil Puria. "Inertial bone conduction: symmetric and anti-symmetric components". In: *Journal of the Association for Research in Otolaryngology* 12.3 (2011), pp. 261–279.
4. Christopher M Williams and Eric M Kramer. "The advantages of a tapered whisker". In: *PLoS One* 5.1 (2010), e8806.
5. Yoon Mo Kim. "Modal Analysis of Continuous Structural System with Tapered Cantilevered Members". PhD thesis. Texas A & M University, 2012.
6. M Gatewood. "Virtual principles in aircraft structures". In: vol. 6. Springer Science & Business Media, 2012. Chap. 5.4-5.7, pp. 169–184.
7. A Bazoune. "Effect of tapering on natural frequencies of rotating beams". In: *Shock and Vibration* 14.3 (2007), pp. 169–179.
8. O Ozdemir Ozgumus and MO Kaya. "Flapwise bending vibration analysis of a rotating double-tapered Timoshenko beam". In: *Archive of Applied Mechanics* 78.5 (2008), pp. 379–392.
9. Yinlei Huo and Zhongmin Wang. "Dynamic analysis of a rotating doubletapered cantilever Timoshenko beam". In: *Archive of Applied Mechanics* 86.6 (2016), pp. 1147–1161.
10. MJ Vasile et al. "Scanning probe tip geometry optimized for metrology by focused ion beam ion milling". In: *Journal of Vacuum Science & Technology B: Microelectronics and Nanometer Structures Processing, Measurement, and Phenomena* 9.6 (1991), pp. 3569–3572.
11. Steven Joseph Gross. "Micromachined switches and cantilever actuators based on piezoelectric lead zirconate titanate (PZT)". PhD thesis. The Pennsylvania State University, Department of Electrical Engineering, 2004.
12. Chieh-Hsiao Wang and Robert J Weber. "Tapered-width micro-cantilever for pull-in voltage control". In: *MEMS Components and Applications for Industry, Automobiles, Aerospace, and Communication II*. Vol. 4981. International Society for Optics and Photonics. 2003, pp. 235–243.
13. Frank Goldschmidtboeing and Peter Woias. "Characterization of different beam shapes for piezoelectric energy harvesting". In: *Journal of micromechanics and microengineering* 18.10 (2008), p. 104013.
14. Jessy Baker, Shad Roundy, and Paul Wright. "Alternative geometries for increasing power density in vibration energy scavenging for wireless sensor networks". In: *3rd international energy conversion engineering conference*. 2005, p. 5617.
15. Wajih U. Syed et al. "Numerical Modeling and Validation of Squeezed-film Damping in Vacuum-packaged Industrial MEMS". In: *Journal of Micromechanics and Microengineering* (2017).
16. Shad Roundy et al. "Improving power output for vibration-based energy scavengers". In: *IEEE Pervasive computing* 4.1 (2005), pp. 28–36.
17. Loreto Mateu and Francesc Moll. "Optimum piezoelectric bending beam structures for energy harvesting using shoe inserts". In: *Journal of Intelligent Material Systems and Structures* 16.10 (2005), pp. 835–845.
18. William Weaver Jr, Stephen P Timoshenko, and Donovan Harold Young. *Vibration problems in engineering*. John Wiley & Sons, 1990.
19. Noël Eduard Du Toit, Brian L Wardle, and Sang-Gook Kim. "Design considerations for MEMS-scale piezoelectric mechanical vibration energy harvesters". In: *Integrated Ferroelectrics* 71.1 (2005), pp. 121–160.
20. Alper Erturk. "Electromechanical modeling of piezoelectric energy harvesters". PhD thesis. Virginia Polytechnic Institute and State University, 2009.

21. Alper Erturk and Daniel J Inman. "On mechanical modeling of cantilevered piezoelectric vibration energy harvesters". In: *Journal of Intelligent Material Systems and Structures* (2008).
22. "Han-Chung" Wang. "Generalized hypergeometric function solutions on the transverse vibration of a class of nonuniform beams". In: *Journal of Applied Mechanics* (1967), pp. 702–708.
23. Dumitru I Caruntu. "Eigenvalue singular problem of factorized fourth-order self-adjoint differential equations". In: *Applied Mathematics and Computation* 224 (2013), pp. 603–610.
24. HH Mabie and CB Rogers. "Transverse vibrations of tapered cantilever beams with end support". In: *The Journal of the Acoustical Society of America* 44.6 (1968), pp. 1739–1741.
25. HH Mabie and CB Rogers. "Transverse vibrations of double-tapered cantilever beams". In: *The Journal of the Acoustical Society of America* 51.5B (1972), pp. 1771–1774.
26. CY Wang. "Vibration of a tapered cantilever of constant thickness and linearly tapered width". In: *Archive of Applied Mechanics* 83.1 (2013), pp. 171–176.
27. Jong-Shyong Wu. *Analytical and numerical methods for vibration analyses*. John Wiley & Sons, 2013.
28. Reza Attarnejad, Negin Manavi, and Alireza Farsad. "Exact solution for the free vibration of a tapered beam with elastic end rotational restraints". In: *Computational Methods*. Springer, 2006, pp. 1993–2003.
29. A Nikoliac and S Saliniac. "Natural frequencies of a tapered cantilever beam of constant thickness and linearly tapered width". In: *The Eight Triennial International Conference HEAVY MACHINERY HM*. 2014, pp. 61–69.
30. Dawid Cekus. "Free vibration of a cantilever tapered Timoshenko beam". In: *Scientific Research of the Institute of Mathematics and Computer Science* 11.4 (2012), pp. 11–17.
31. Serge Abrate. "Vibration of non-uniform rods and beams". In: *Journal of sound and vibration* 185.4 (1995), pp. 703–716.
32. Sajal Sagar Singh, Prem Pal, and Ashok Kumar Pandey. "Pull-in analysis of non-uniform microcantilever beams under large deflection". In: *Journal of Applied Physics* 118.20 (2015), p. 204303.
33. Qibo Mao and Stanislaw Pietrzko. "Free vibration analysis of a type of tapered beams by using Adomian decomposition method". In: *Applied Mathematics and Computation* 219.6 (2012), pp. 3264–3271.
34. Ozge Ozdemir Ozgumus and Metin O Kaya. "Energy expressions and free vibration analysis of a rotating double tapered Timoshenko beam featuring bending-torsion coupling". In: *International journal of engineering science* 45.2 (2007), pp. 562–586.
35. Mashinchi Joubari et al. "Determination of Periodic Solution for Tapered Beams with Modified Iteration Perturbation Method". In: *Journal of Applied and Computational Mechanics* 1.1 (2014), pp. 44–51.
36. Safa Bozkurt Coskun, Mehmet Tarik Atay, and Baki Ozturk. "Transverse vibration analysis of Euler-Bernoulli beams using analytical approximate techniques". In: *Advances in Vibration Analysis Research*. InTech, 2011.
37. Mahmoud Bayat, Iman Pakar, and Mahdi Bayat. "Analytical study on the vibration frequencies of tapered beams". In: *Latin American Journal of Solids and Structures* 8.2 (2011), pp. 149–162.
38. CW Bert and M Malik. "Free vibration analysis of tapered rectangular plates by differential quadrature method: a semi-analytical approach". In: *Journal of Sound and Vibration* 190.1 (1996), pp. 41–63.
39. Shu-qi Guo and Shao-pu Yang. "Transverse vibrations of arbitrary nonuniform beams." In: *Applied Mathematics & Mechanics* 35.5 (2014).
40. Jung-Chang Hsu, Hsin-Yi Lai, and Cha'o-Kuang Chen. "Free vibration of non-uniform Euler–Bernoulli beams with general elastically end constraints using Adomian modified decomposition method". In: *Journal of Sound and Vibration* 318.4-5 (2008), pp. 965–981.
41. S Naguleswaran. "Vibration of an Euler-Bernoulli beam of constant depth and with linearly varying breadth". In: *Journal of sound and vibration* 153.3 (1992), pp. 509–522.
42. *MEMS+ Users Guide*. Coventor Inc. 2002.

43. Yaowen Yang and Lihua Tang. "Equivalent circuit modeling of piezoelectric energy harvesters". In: *Journal of Intelligent Material Systems and Structures* (2009), pp. 2223–2235.
44. Nan Wang et al. "AlN wideband energy harvesters with wafer-level vacuum packaging utilizing three-wafer bonding". In: *Micro Electro Mechanical Systems (MEMS), 2017 IEEE 30th International Conference on*. IEEE. 2017, pp. 841–844.
45. CB Williams and Rob B Yates. "Analysis of a micro-electric generator for microsystems". In: *Sensors and Actuators A: Physical* 52.1-3 (1996), pp. 8–11.
46. Wajih U. Syed et al. "Numerical Modeling and Validation of Squeezed-film Damping in Vacuum-packaged Industrial MEMS". In: *Journal of Micromechanics and Microengineering* (2017).

Chapter 4
Modal Analysis

The first step in developing the semi-analytical model of a tapered piezoelectric energy harvester is to formulate a structural model for *modal analysis* of a multilayered, prismatic, tapered cantilever structure. The structure modeled here is strictly prismatic, assuming straight cuts in the structure fabrication process without undercutting. When undercutting is considered, the model can be updated to consider a taper along the structure's thickness. The governing equation of motion for a tapered cantilever is first set up before formulating the spatial modal equation via the separation of variables. The modal equation is then perturbed over the taper of the beam, and the resulting differential eigenvalue problem is solved using a carefully derived Green's function with the help of a symbolic library programmed in Mathematica. The resonance frequency and mode shapes of the resulting semi-analytical model are validated against FEM modal analysis results. The results of the validation and error analysis are presented after discussing the symbolic modeling framework in Chap. 5.

In this chapter, we present a novel semi-analytical model of the vibration dynamics of a multilayered tapered prismatic cantilever. The derivation of the model is based on a perturbation expansion of the tapered cantilever's partial differential equations using the Euler–Bernoulli beam theory as a starting point. The proposed semi-analytical solution of the model enables a 101-sample modal analysis to run more than 250× faster when using 20 perturbation components than a computationally efficient commercial FEM solution that uses 101 shell-type finite elements along the beam length and 6 elements along the beam width. In this case, the prediction error, in the resonance frequency, of the primary mode model is less than 0. 1% for 20 perturbation components and less than 10% for 4 components.

W. U. Syed, I. (Abe) M. Elfadel, *Tapered Beams in MEMS*,
https://doi.org/10.1007/978-3-031-66391-8_4

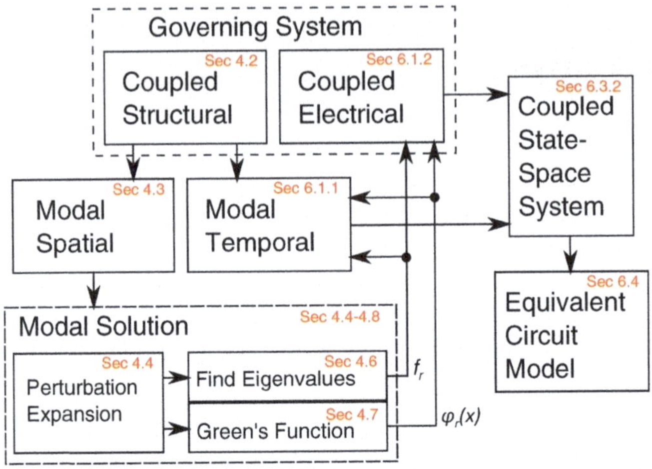

Fig. 4.1 Tapered PEH system formulation

4.1 Tapered Beam EH Model Formulation Workflow

Figure 4.1 explains the workflow for formulating and solving the semi-analytical model of a tapered beam energy harvester which starts from the development of the governing system of equations and ends with the formulation of the electromechanical coupled state-space system and the equivalent circuit model (ECM) of a tapered piezoelectric energy harvester. Steady-state or transient analysis can subsequently be performed on the state-space system and the ECM. Chapter 4 illustrates the development of an algorithm to solve for the resonance frequencies and mode shapes of a tapered cantilever. Chapter 5 describes the symbolic computation tools used to develop the algorithm. Chapter 6 uses these results of the modal analysis to build the electromechanical system of the tapered PEH with piezoelectric coupling.

Following the Euler–Bernoulli methodology, force and moment balancing, in conjunction with coupling models to the system external to the physical structure of the tapered beam, help derive the coupled structural equation as the governing equation of motion in Sect. 4.2. We apply separation of variables on the governing equation of motion to obtain the spatial modal equation in Sect. 4.3, which we solve for the resonance frequency and mode shapes to complete the modal analysis. However, we discover that a differential eigenanalysis of the tapered cantilever modal equation is not straightforward. We formulate a perturbation analysis of the modal equation of the tapered beam around the solution of the rectangular beam in Sect. 4.4. We solve the perturbation system of the differential eigenvalue problem in the subsequent sections, including the derivation of an eigenvalue solution in Sect. 4.7 and the solution of Green's function of the iterative perturbation analysis system in Sect. 4.7. The complete modal analysis strategy is shown in Algorithm 4.1.

The separation of variables performed in Sect. 4.3 also results in the dynamic modal response equation derived in Sect. 6.1.2 in Chap. 6. In the end, in Sect. 6.1.3, the electromechanically coupled circuit equation is set up by balancing the total current entering a node in the circuit with a resistive load across the EH terminals. An electromechanically coupled state-space system is then built from the modal response equation and the circuit equation in Sect. 6.3.2.

4.2 Governing Equation of Motion

The first step in deriving the governing equation of motion as shown in Fig. 4.2 is to formulate the structural equation of a cantilever beam that is coupled with piezoelectric physics via the voltage across the piezoelectric layer $v(t)$. The governing equation of the structural system can be coupled to external physics utilizing the coupling factor $\theta(x)$.

The geometry of a tapered multilayered cantilever used to develop the TPEH model is shown in Fig. 4.3. This structure is shown to be constructed with a substructure layer and a piezoelectric layer, both with different structural properties. To evaluate the structure mechanics in this chapter, we will consider a more generic structure with an arbitrary number, N_l, of layers shown in Fig. 4.4. Given this layered structure, the piezoelectric layer can then be arbitrarily placed at layer p.

The width of the beam at the fixed end is b_0 and at the free end a distance L away from the anchor is b_L. The total thickness of the structure is t, whereas the layer number indexes the thickness of each layer. Figure 4.4 illustrates a beam with N_L layers of arbitrary thicknesses t_i, $i \in 1, \cdots, N_L$.

The governing equation of motion for such a tapered Euler–Bernoulli beam when subjected to viscous and strain rate damping can be obtained using Newtonian or

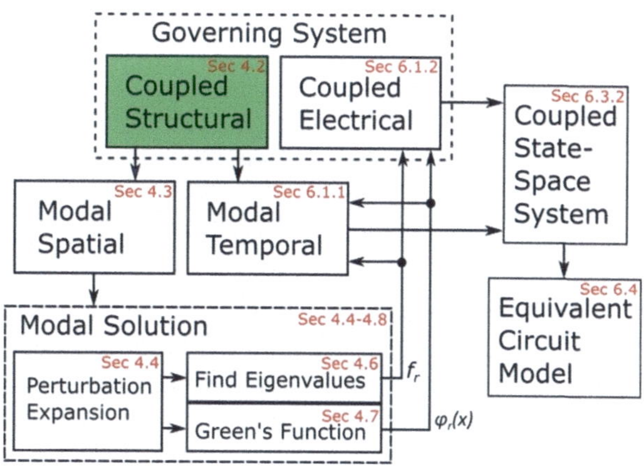

Fig. 4.2 Tapered PEH system formulation: finding the coupled structural governing equation

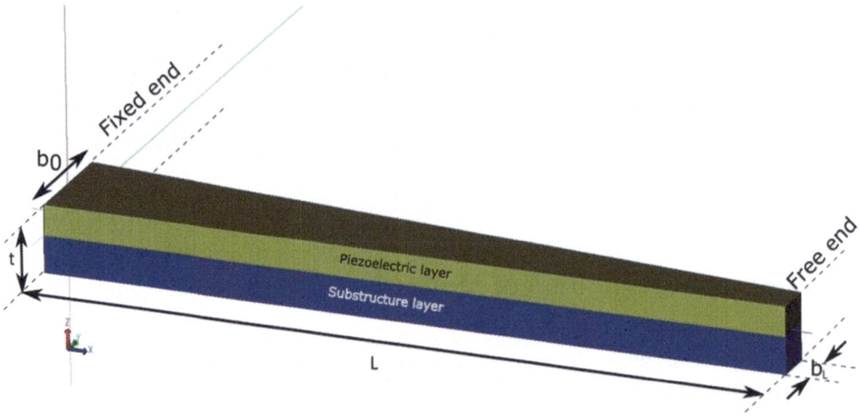

Fig. 4.3 Tapered PEH geometry

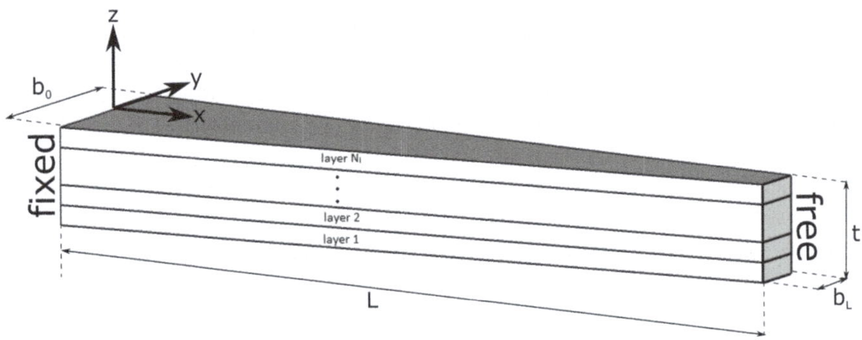

Fig. 4.4 Tapered cantilever geometry

Hamiltonian approaches [1]. The governing equation of the tapered PEH can be set using force balancing as [2]

$$\frac{\partial^2 M(x,t)}{\partial x^2} + c_a \frac{\partial w_{rel}(x,t)}{\partial t} + m(x) \frac{\partial^2 w_{rel}(x,t)}{\partial t^2} = - m(x) \frac{\partial^2 w_b(x,t)}{\partial t^2} - c_a \frac{\partial w_b(x,t)}{\partial t}, \tag{4.1}$$

where $M(x,t)$ is the internal bending moment, c_a is the viscous air damping coefficient, $m(x)$ is the mass per unit length of the beam at x, and $w_{rel}(x,t)$ is the transverse beam deflection relative to its base, whereas $w_b(x,t) = \gamma_g(x)g(t) + \gamma_h(x)h(t)$ is the base motion with $g(t)$ being the translational motion and $h(t)$ the angular motion and $\gamma_g(x)$ and $\gamma_h(x)$ the corresponding displacement functions. For a cantilever beam $\gamma_g(x) = 1$ and $\gamma_h(x) = x$ such that $w_b(x,t) = g(t) + x\,h(t)$.

4.2.1 Internal Moment

Since we are interested in the modal analysis of the structure, we ignore the cross-physics coupling and assume a short circuit across the external physics port thus making ours a one-port system. Also, we only consider the undamped internal moment, $M_u(x, t)$, as the net moment ignoring damping. We then have

$$M(x,t) = M_u(x,t) = -\sum_{i=1}^{N_L} \int_{z_{i-1}}^{z_i} T_i(x,t)b(x)z dz, \tag{4.2}$$

where z_i is the z-location of the ith layer relative to the neutral axis and $b(x)$ or b_x is the width of the beam at distance x from the base. The stress–strain relationship for the ith linear elastic layer is $T_i = Y_i S_i$, where T is the stress, S is the strain, and Y is Young's modulus of the material. Note that in the case of a piezoelectric layer, the stress–strain relationship for that layer is modified using the piezoelectric constitutive equations as [3]

$$T_p = Y_p(S_p - d_{31}E_3),$$

where T is the stress, S is the strain, Y is Young's modulus, d is the piezoelectric constant, and E is the electric field. The subscript numbers represent directions. The bending strain $S(x)$ is defined for small deflections in terms of radius of curvature as

$$S(x,z,t) = -z\frac{\partial^2 w_{rel}(x,t)}{\partial x^2}. \tag{4.3}$$

Using bending strain for the structural and piezoelectric layers in Eq. (4.2), the undamped internal moment can be written as

$$M_u(x,t) = \frac{\partial^2 w_{rel}(x,t)}{\partial x^2} \sum_{i=1}^{N} \left(Y_i b(x) \int_{z_i}^{z_{i+1}} z^2 dz \right) - v(t) Y_p b(x) \frac{d_{31}}{h_p} \int_{z_p}^{z_{p+1}} z dz,$$

where p is the index of the piezoelectric layer and $v(t) = E_3 h_p$ is the voltage across the piezoelectric layer. Using $YI_i(x) = Y_i b(x) \int_{z_i - z_n}^{z_{i+1} - z_n} z^2 dz$ and $YI(x) = \sum_{i=1}^{N} YI_i(x)$, the above equation simplifies to

$$M_u(x,t) = YI(x)\frac{\partial^2 w_{rel}(x,t)}{\partial x^2} + \theta(x)v(t), \tag{4.4}$$

where the coupling term $\theta(x)$ is

$$\theta(x) = -Y_p d_{31} b(x) z_{pc},$$

where z_{pc} is the distance of the center of the piezoelectric layer from the neutral axis. Considering the contribution of strain rate damping to the internal moment, Eq. (4.4) can be adjusted as

$$M(x,t) = YI(x)\frac{\partial^2 w_{rel}(x,t)}{\partial x^2} + \theta(x)v(t) + c_s I(x)\frac{\partial^3 w_{rel}(x,t)}{\partial x^2 \partial t}, \tag{4.5}$$

where $c_s I(x)$ is the equivalent coefficient at location x of the structural viscoelasticity term representing strain rate damping with c_s, the equivalent strain rate damping coefficient, and $I(x)$, the homogenized moment of inertia of the beam cross section at x. Allowing for electrodes of arbitrary length starting from the fixed end, Heaviside functions are used to define the extent of electrodes over the beam to a point L_e distance from the fixed end. Equation (4.5) can then be rewritten as

$$\begin{aligned} M(x,t) = YI(x)\frac{\partial^2 w_{rel}(x,t)}{\partial x^2} &+ \theta(x)v(t)\left(H(x) - H(x-L_e)\right) \\ + c_s I(x)\frac{\partial^3 w_{rel}(x,t)}{\partial x^2 \partial t}&. \end{aligned} \tag{4.6}$$

Identifying the structural mechanical bending moment $M_m(x,t)$ as

$$M_m(x,t) = YI(x)\frac{\partial^2 w_{rel}(x,t)}{\partial x^2},$$

the piezoelectric coupling moment $M_e(x,t)$ as

$$M_e(x,t) = \theta(x)v(t)\left(H(x) - H(x-L_e)\right),$$

and the strain rate damping bending moment $M_d(x,t)$ as

$$M_d(x,t) = c_s I(x)\frac{\partial^3 w_{rel}(x,t)}{\partial x^2 \partial t},$$

in Eq. (4.6), the net damping moment, $M(x,t)$ can be written as

$$M(x,t) = M_m(x,t) + M_e(x,t) + M_d(x,t).$$

Next, the second derivatives of each of the three components of $M(x,t)$ are taken with respect to x before substituting them in Eq. (4.1).

4.2.2 Applying Equilibrium Condition

The force density caused by the transverse mechanical bending moment is evaluated by taking the second derivative of $M_m(x,t)$ with respect to x. Using product rule gives

$$\begin{aligned}\frac{\partial^2 M_m(x,t)}{\partial x^2} =& YI(x)\frac{\partial^4 w_{rel}(x,t)}{\partial x^4} + \frac{\partial^2 YI(x)}{\partial x^2}\frac{\partial^2 w_{rel}(x,t)}{\partial x^2} \\ &+ 2\frac{\partial YI(x)}{\partial x}\frac{\partial^3 w_{rel}(x,t)}{\partial x^3}.\end{aligned} \tag{4.7}$$

The force density from the piezoelectric bending moment is similarly derived by using the product rule to find the double derivative of $M_e(x,t)$ with respect to x. Noting that the derivative of the Heaviside function is the Dirac delta function, $\delta(x)$, the force density from piezoelectric bending moment is evaluated as

$$\begin{aligned}\frac{\partial^2 M_e(x,t)}{\partial x^2} =& v(t)\left(\frac{d^2\theta(x)}{dx^2}\left(H(x) - H(x-L_e)\right)\right. \\ &\left.+\theta(x)\left(\frac{d\delta(x)}{dx} - \frac{d\delta(x-L_e)}{dx}\right) + 2\frac{d\theta(x)}{dx}\left(\delta(x) - \delta(x-L_e)\right)\right).\end{aligned} \tag{4.8}$$

The strain rate damping induces a force density that is similarly evaluated as

$$\begin{aligned}\frac{\partial^2 M_d(x,t)}{\partial x^2} =& c_s I(x)\frac{\partial^5 w_{rel}(x,t)}{\partial x^4 \partial t} + c_s\frac{d^2 I(x)}{dx^2}\frac{\partial^3 w_{rel}(x,t)}{\partial x^2 \partial t} \\ &+ 2c_s\frac{dI(x)}{dx}\frac{\partial^4 w_{rel}(x,t)}{\partial x^3 \partial t}.\end{aligned} \tag{4.9}$$

The width of a triangular beam $b(x)$ varies linearly with x as $b(x) = b_0 + \frac{x}{L}(b_L - b_0)$. Defining α, the tapering factor as

$$\alpha = \frac{b_0 - b_L}{b_0},$$

such that we can rewrite the linearly tapering width of the tapered beam as

$$b(x) = b_0\left(1 - \alpha\frac{x}{L}\right),$$

which leads to a linear profile for various beam properties such as the second moment of inertia, $I(x) = I_0(1-\alpha\frac{x}{L})$; the flexural rigidity, $YI(x) = YI_0(1-\alpha\frac{x}{L})$; and the mass per unit length $m(x) = m_0(1-\alpha\frac{x}{L})$. Also, the piezoelectric coupling term $\theta(x)$ can be written as $\theta(x) = \theta_0(1-\alpha\frac{x}{L})$, where $\theta_0 = -Y_p d_{31} b_0 z_{pc}$. The first derivatives of these *linear* terms reduce to a constant and their higher derivatives vanish. The force density terms in Eqs. (4.7), (4.9), and (4.8) are reduced in terms of α to

$$\frac{\partial^2 M_m(x,t)}{\partial x^2} = YI_0\left(1-\alpha\frac{x}{L}\right)\frac{\partial^4 w_{rel}(x,t)}{\partial x^4} - 2YI_0\frac{\alpha}{L}\frac{\partial^3 w_{rel}(x,t)}{\partial x^3},$$

$$\frac{\partial^2 M_d(x,t)}{\partial x^2} = c_s I_0\left(1-\alpha\frac{x}{L}\right)\frac{\partial^5 w_{rel}(x,t)}{\partial x^4\partial t} - 2c_s I_0\frac{\alpha}{L}\frac{\partial^4 w_{rel}(x,t)}{\partial x^3\partial t}, \text{ and}$$

$$\begin{aligned}\frac{\partial^2 M_e(x,t)}{\partial x^2} =& v(t)\left(\theta_0\left(1-\alpha\frac{x}{L}\right)\left(\frac{d\delta(x)}{dx} - \frac{d\delta(x-L_e)}{dx}\right)\right.\\ &\left.-2\theta_0\frac{\alpha}{L}\left(\delta(x)-\delta(x-L_e)\right)\right).\end{aligned}$$

Putting the force density components together and rearranging, the total force density can be written as

$$\begin{aligned}\frac{\partial^2 M(x,t)}{\partial x^2} =& YI_0\frac{\partial^4 w_{rel}(x,t)}{\partial x^4} + c_s I_0\frac{\partial^5 w_{rel}(x,t)}{\partial x^4\partial t} + \theta_0 v(t)\left(\frac{d\delta(x)}{dx} - \frac{d\delta(x-l)}{dx}\right)\\ &+\alpha\left[-\left(\frac{2YI_0}{L}\frac{\partial^3 w_{rel}(x,t)}{\partial x^3} + \frac{2c_s I_0}{L}\frac{\partial^4 w_{rel}(x,t)}{\partial x^3\partial t} + 2\frac{\theta_0}{L}v(t)\left(\delta(x)-\delta(x-L_e)\right)\right)\right.\\ &\left.-x\left(\frac{YI_0}{L}\frac{\partial^4 w_{rel}(x,t)}{\partial x^4} + \frac{c_s I_0}{L}\frac{\partial^5 w_{rel}(x,t)}{\partial x^4\partial t} + \frac{\theta_0}{L}v(t)\left(\frac{d\delta(x)}{dx} - \frac{d\delta(x-L_e)}{dx}\right)\right)\right].\end{aligned}$$

Substituting the force density into the structural governing equation in Eq. (4.1) and rearranging, we get

$$\begin{aligned}&YI_0\frac{\partial^4 w_{rel}(x,t)}{\partial x^4} + c_s I_0\frac{\partial^5 w_{rel}(x,t)}{\partial x^4\partial t} + \theta_0 v(t)\left(\frac{d\delta(x)}{dx} - \frac{d\delta(x-L_e)}{dx}\right)\\ &+c_a\frac{\partial w_{rel}(x,t)}{\partial t} + c_a\frac{dg(t)}{dt} + c_a x\frac{dh(t)}{dt} + m_0\frac{\partial^2 w_{rel}(x,t)}{\partial t^2} + m_0\frac{d^2 g(t)}{dt^2}\\ &+\alpha\left(-\left(\frac{2YI_0}{L}\frac{\partial^3 w_{rel}(x,t)}{\partial x^3} + \frac{2c_s I_0}{L}\frac{\partial^4 w_{rel}(x,t)}{\partial x^3\partial t}\right.\right.\\ &\left.+2\frac{\theta_0}{L}v(t)\left(\delta(x)-\delta(x-L_e)\right)\right) - x\left(\frac{YI_0}{L}\frac{\partial^4 w_{rel}(x,t)}{\partial x^4}\right.\end{aligned}$$

(continued)

$$+\frac{m_0}{L}\frac{\partial^2 w_{rel}(x,t)}{\partial t^2}+\frac{m_0}{L}\frac{d^2 g(t)}{dt^2}+\frac{m_0}{L}x\frac{d^2 h(t)}{dt^2}+\frac{c_s I_0}{L}\frac{\partial^5 w_{rel}(x,t)}{\partial x^4 \partial t}$$
$$+\frac{\theta_0}{L}v(t)\left(\frac{d\delta(x)}{dx}-\frac{d\delta(x-L_e)}{dx}\right)\Bigg)\Bigg)=0. \tag{4.10}$$

Choosing the tapering factor α as the taper control parameter, we notice in Eq. (4.10) that when $\alpha = 0$, Eq. (4.10) reduces to the rectangular beam case with a width of b_0 derived by Erturk et al. [4] and shown in Chap. 2.

4.3 Spatial Modal Equation Formulation

Once we have formulated the electromechanically coupled structural governing equation, we look for ways to solve it analytically. Given that Eq. (4.10) and the boundary conditions for a fixed-free beam are linear and homogeneous, we can use the separation of variables for deriving separate spatial and temporal equations. The first step in the separation of variables is the formulation of the spatial equations as highlighted in Fig. 4.5.

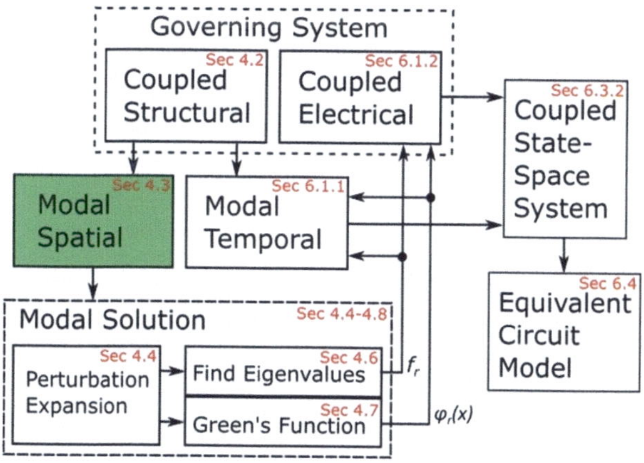

Fig. 4.5 Tapered PEH system formulation: spatial modal equation

4.3.1 Separation of Variables

Using separation of variables we replace $w_{rel}(x,t)$ with its modal summation $w_{rel}(x,t) = \sum_{r=1}^{\infty} \eta_r(t)\phi_r(x)$. For the rth mode, we substitute $w_{rel}(x,t) = \eta_r(t)\phi_r(x)$ in Eq. (4.10) to get

$$\begin{aligned}
& YI_0\eta_r(t)\frac{d^4\phi_r(x)}{dx^4} + c_sI_0\frac{d^4\phi_r(x)}{dx^4}\frac{d\eta_r(t)}{dt} + \theta_0 v(t)\left(\frac{d\delta(x)}{dx} - \frac{d\delta(x-L_e)}{dx}\right) \\
& + c_a\phi_r(x)\frac{d\eta_r(t)}{dt} + c_a\frac{dg(t)}{dt} + c_a x\frac{dh(t)}{dt} + m_0\phi_r(x)\frac{d^2\eta_r(t)}{dt^2} + m_0\frac{d^2g(t)}{dt^2} \\
& + m_0 x\frac{d^2h(t)}{dt^2} + \alpha\left(-\left(\frac{2YI_0}{L}\eta_r(t)\frac{d^3\phi_r(x)}{dx^3} + \frac{2c_sI_0}{L}\frac{d^3\phi_r(x)}{dx^3}\frac{d\eta_r(t)}{dt}\right.\right. \\
& \left. +2\frac{\theta_0}{L}v(t)\left(\delta(x) - \delta(x-L_e)\right)\right) - x\left(\frac{YI_0}{L}\eta_r(t)\frac{d^4\phi_r(x)}{dx^4} + \frac{m_0}{L}\phi_r(x)\frac{d^2\eta_r(t)}{dt^2}\right. \\
& + \frac{m_0}{L}\frac{d^2g(t)}{dt^2} + \frac{m_0}{L}x\frac{d^2h(t)}{dt^2} + \frac{c_sI_0}{L}\frac{d^4\phi_r(x)}{dx^4}\frac{d\eta_r(t)}{dt} \\
& \left.\left. + \frac{\theta_0}{L}v(t)\left(\frac{d\delta(x)}{dx} - \frac{d\delta(x-L_e)}{dx}\right)\right)\right) = 0.
\end{aligned} \tag{4.11}$$

4.3.2 Spatial Modal Equation of the Free Undamped Beam

Since we are interested in the modal analysis of the structure, we ignore the cross-physics coupling and assume a short circuit across the external physics port thus making ours a one-port system. Also, we only consider the undamped internal moment, $M_u(x,t)$ as the net moment ignoring damping.

For a free undamped beam, setting the forcing and the damping terms to zero gives

$$\begin{aligned}
& YI_0\eta_r(t)\frac{d^4\phi_r(x)}{dx^4} + \theta_0 v(t)\left(\frac{d\delta(x)}{dx} - \frac{d\delta(x-L_e)}{dx}\right) + m_0\phi_r(x)\frac{d^2\eta_r(t)}{dt^2} \\
& + \alpha\left(-\left(\frac{2YI_0}{L}\eta_r(t)\frac{d^3\phi_r(x)}{dx^3} + 2\frac{\theta_0}{L}v(t)\left(\delta(x) - \delta(x-L_e)\right)\right)\right. \\
& - x\left(\frac{YI_0}{L}\eta_r(t)\frac{d^4\phi_r(x)}{dx^4} + \frac{m_0}{L}\phi_r(x)\frac{d^2\eta_r(t)}{dt^2}\right. \\
& \left.\left. + \frac{\theta_0}{L}v(t)\left(\frac{d\delta(x)}{dx} - \frac{d\delta(x-L_e)}{dx}\right)\right)\right) = 0.
\end{aligned}$$

We ignore piezoelectric coupling via setting $v(t) = 0$ for short-circuit analysis to get

$$YI_0 \frac{1}{\phi_r(x)} \frac{d^4\phi_r(x)}{dx^4} + m_0 \frac{1}{\eta_r(t)} \frac{d^2\eta_r(t)}{dt^2} + \alpha \left(- \left(\frac{2YI_0}{L} \frac{1}{\phi_r(x)} \frac{d^3\phi_r(x)}{dx^3} \right) \right.$$
$$\left. - x \left(\frac{YI_0}{L} \frac{1}{\phi_r(x)} \frac{d^4\phi_r(x)}{dx^4} + \frac{m_0}{L} \frac{1}{\eta_r(t)} \frac{d^2\eta_r(t)}{dt^2} \right) \right) = 0.$$

Separating the spatial and temporal terms, we get

$$\frac{1}{\frac{m_0}{YI_0}\left(1 - \alpha \frac{x}{L}\right)} \left(\frac{1}{\phi_r(x)} \frac{d^4\phi_r(x)}{dx^4} + \alpha \left(- \left(\frac{2}{L} \frac{1}{\phi_r(x)} \frac{d^3\phi_r(x)}{dx^3} \right) \right. \right.$$
$$\left. \left. - x \left(\frac{1}{L} \frac{1}{\phi_r(x)} \frac{d^4\phi_r(x)}{dx^4} \right) \right) \right) = - \frac{1}{\eta_r(t)} \frac{d^2\eta_r(t)}{dt^2}.$$

The equality between the spatial and temporal terms on the two sides of the equation implies that they should both be equal to a constant. The temporal half of the equation imposes this constant to be positive which we choose to be ω_r^2 for mode r to give

$$- \frac{YI_0}{m_0} \frac{1}{\left(1 - \alpha \frac{x}{L}\right)} \left(\frac{1}{\phi_r(x)} \frac{d^4\phi_r(x)}{dx^4} + \alpha \left(- \left(\frac{2}{L} \frac{1}{\phi_r(x)} \frac{d^3\phi_r(x)}{dx^3} \right) \right. \right.$$
$$\left. \left. - x \left(\frac{1}{L} \frac{1}{\phi_r(x)} \frac{d^4\phi_r(x)}{dx^4} \right) \right) \right) = -\omega_r^2.$$

Let $\beta_r = \frac{m_0 \omega_r^2}{YI_0}$ be the eigenvalue for mode r. Then after rearranging, we get

$$\frac{d^4\phi_r(x)}{dx^4} - \beta_r \phi_r(x) + \alpha \left(-\frac{2}{L} \frac{d^3\phi_r(x)}{dx^3} - \frac{x}{L} \frac{d^4\phi_r(x)}{dx^4} + \beta_r \frac{x}{L} \phi_r(x) \right) = 0. \tag{4.12}$$

Equation (4.12) is the *mode-shape equation of the tapered beam* and we will need to solve it for $\phi_r(x)$ and β_r to complete the modal analysis of the beam. We do not expect the trial functions of a rectangular beam to work in this case. It turns out that the solution to this equation is in terms of some special functions which are not very easy to deal with once the boundary conditions are applied to eliminate the degrees of freedom in the solution.

4.3.2.1 Modal Solution in Terms of Special Functions

It turns out that the solution to this equation is in terms of some special series functions [5] which are not very easy to deal with once the boundary conditions are applied to eliminate the degrees of freedom in the solution. A quick query in Mathematica returns the mode-shape solution to Eq. (4.12) before imposing the boundary conditions as

$$\begin{aligned}\phi_r(x) =& \frac{c_2\,(-\beta_r)^{\frac{1}{4}}\,(l-\alpha x)\,{}_0F_3\left(\{\}; \frac{3}{4}, 1, \frac{5}{4}; \frac{(l-x\alpha)^4\beta_r}{256\alpha^4}\right)}{4\alpha} \\ &+ c_4\,{}_0F_3\left(\{\}; \frac{1}{2}, \frac{3}{4}, \frac{3}{4}; \frac{(l-x\alpha)^4\beta_r}{256\alpha^4}\right) \\ &+ \frac{c_1\,(-\beta_r)^{\frac{1}{2}}\,(l-\alpha x)^2\,{}_0F_3\left(\{\}; \frac{5}{4}, \frac{5}{4}, \frac{3}{2}; \frac{(l-x\alpha)^4\beta_r}{256\alpha^4}\right)}{16\alpha^2} \\ &+ c_3 G_{0,4}^{2,0}\left(\frac{(l-x\alpha)^4\beta_r}{256\alpha^4} \middle| \frac{1}{4}, \frac{1}{4}, 0, \frac{1}{2}\right),\end{aligned}$$

where ${}_pF_q(a_1\cdots a_p; b_1\cdots b_q; z)$ is the generalized hypergeometric function and $G_{p,q}^{m,n}\left(z\,\middle|\,\begin{matrix}a_1,\cdots,a_p\\ b_1,\cdots,b_q\end{matrix}\right)$ is the Meijer-G function. The empty sets in the first argument of hypergeometric functions follow from $p=0$ in ${}_0F_q(\{\},\cdot,\cdot)$ instances in the solution. The constants c_1, c_2, c_3, and c_4 are determined using the boundary conditions for the type of beam being solved for. The boundary conditions for the mode shapes of a cantilever are given in Eq. (2.14). For $\beta_r > 0$, which is what we would expect from the analogy of the rectangular beam, the above solution results in complex-valued mode shapes. Moreover, the application of the boundary conditions leads to a rather cumbersome, multi-term expression involving the two special functions, which becomes quite intractable and impractical for use with analysis of the beam for design purposes. We, therefore, turn our attention to an approximate solution of the mode shapes of a tapered beam.

4.4 Perturbation Expansion of the Modal Solution

The modal equation of a tapered beam is shown in Eq. (4.12) and has the additional third derivative term that changes the nature of the equation from the rectangular beam equation; hence, the solution to it is not straightforward anymore. Inspecting Eq. (4.12) we notice that the taper ratio α naturally sets up the tapered system as a perturbation on the rectangular beam which is the nominal case of the equation when $\alpha = 0$. Since we know the solution of the nominal case [4], we will set up a perturbation analysis for Eq. (4.12) and formulate a perturbation analysis highlighted in Fig. 4.6 before solving it in later sections.

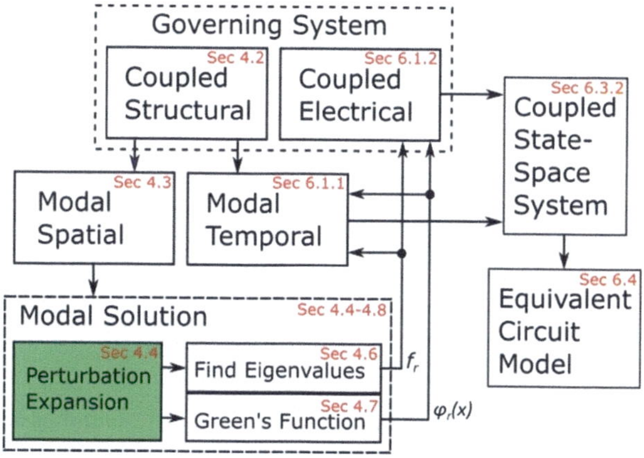

Fig. 4.6 Tapered PEH system formulation: perturbed system for the spatial modal equation

4.4.1 *Iterative Perturbed Tapered System (ITPS) Formulation*

To set up perturbation analysis we make the following substitutions in Eq. (4.12) [6–8]:

$$\phi_r(x) = \sum_{n=0}^{\infty} \psi_{r,n}(x)\alpha^n, \text{ and} \tag{4.13}$$

$$\beta_r = \mu_{r,0} + \sum_{n=1}^{\infty} \mu_{r,n}\alpha^n. \tag{4.14}$$

We will drop the subscript r for the mode number from the perturbation components of the mode shapes, $\psi_n = \psi_{r,n}$, and eigenvalues, $\mu_n = \mu_{r,n}$ for the sake of clarity. The equation obtained will thus have a different solution for each mode r.

$$\begin{aligned}
&\frac{d^4}{dx^4}\left(\sum_{n=0}^{\infty} \psi_n(x)\alpha^n\right) - \left(\mu_0 + \sum_{n=1}^{\infty} \mu_n\alpha^n\right)\sum_{m=0}^{\infty} \psi_m(x)\alpha^m \\
&+\alpha\left(-\frac{2}{L}\frac{d^3}{dx^3}\left(\sum_{n=0}^{\infty} \psi_n(x)\alpha^n\right) - \frac{x}{L}\frac{d^4}{dx^4}\left(\sum_{n=0}^{\infty} \psi_n(x)\alpha^n\right)\right. \\
&\left.+\left(\mu_0 + \sum_{n=1}^{\infty} \mu_n\alpha^n\right)\frac{x}{L}\sum_{m=0}^{\infty} \psi_m(x)\alpha^m\right) = 0.
\end{aligned}$$

Rearranging the summation signs we get

$$\sum_{n=0}^{\infty}\frac{d^4\psi_n(x)}{dx^4}\alpha^n - \mu_0\sum_{n=0}^{\infty}\psi_n(x)\alpha^n = \frac{2}{L}\sum_{n=0}^{\infty}\frac{d^3\psi_n(x)}{dx^3}\alpha^{n+1}$$
$$+\frac{x}{L}\sum_{n=0}^{\infty}\frac{d^4\psi_n(x)}{dx^4}\alpha^{n+1} - \mu_0\frac{x}{L}\sum_{n=0}^{\infty}\psi_n(x)\alpha^{n+1}$$
$$-\frac{x}{L}\sum_{n=1}^{\infty}\sum_{m=0}^{\infty}\mu_n\psi_m(x)\alpha^{n+m+1} + \sum_{n=1}^{\infty}\sum_{m=0}^{\infty}\mu_n\psi_m(x)\alpha^{n+m}.$$

Comparing the powers of α and rearranging we get a differential equation for each of the perturbation components, $\psi_j(x)$, where $0 \le j \le n$. For the terms involving eigenvalue expansions (μ_n) we substitute $m = n - j$ to drop one of the summations when selecting out the $\psi_n(x)$ term to give

$$\frac{d^4\psi_n(x)}{dx^4} - \mu_0\psi_n(x) = \frac{2}{L}\frac{d^3\psi_{n-1}(x)}{dx^3} + \frac{x}{L}\left(\frac{d^4\psi_{n-1}(x)}{dx^4} - \mu_0\psi_{n-1}(x)\right)$$
$$-\frac{x}{L}\sum_{j=1}^{n-1}\mu_j\psi_{n-1-j}(x) + \sum_{j=1}^{n}\mu_j\psi_{n-j}(x).$$

We notice that the term in parentheses is the same as the left-hand side of the equation indexed with $n-1$. We expand this term recursively using the right-hand side terms to get

$$\frac{d^4\psi_n(x)}{dx^4} - \mu_0\psi_n(x) = \frac{2}{L}\frac{d^3\psi_{n-1}(x)}{dx^3} + \frac{x}{L}\left(\frac{2}{L}\frac{d^3\psi_{n-2}(x)}{dx^3}\right.$$
$$\left.+\frac{x}{L}\left(\frac{d^4\psi_{n-2}(x)}{dx^4} - \mu_0\psi_{n-2}(x)\right) - \frac{x}{L}\sum_{j=1}^{n-2}\mu_j\psi_{n-2-j}(x) + \sum_{j=1}^{n-1}\mu_j\psi_{n-1-j}(x)\right)$$
$$-\frac{x}{L}\sum_{j=1}^{n-1}\mu_j\psi_{n-1-j}(x) + \sum_{j=1}^{n}\mu_j\psi_{n-j}(x).$$

We collect the third derivative terms that appear in the recursive expansion to form a power series. We also notice that the last term in this recursion $\frac{d^4\psi_0(x)}{dx^4} - \mu_0\psi_0(x)$ is equal to 0 and hence the recursion is gracefully terminated. We also collect the eigenvalue terms to construct two power series terms. Also, the second and third summations should terminate at $n-1$ to give

$$\frac{d^4\psi_n(x)}{dx^4} - \mu_0\psi_n(x) = \frac{2}{L}\sum_{k=0}^{n-1}\left(\frac{x}{L}\right)^k \frac{d^3\psi_{n-1-k}(x)}{dx^3}$$
$$-\sum_{k=1}^{n-1}\left(\frac{x}{L}\right)^k \sum_{j=1}^{n-k}\mu_j\psi_{n-k-j}(x) + \sum_{k=0}^{n-1}\left(\frac{x}{L}\right)^k \sum_{j=1}^{n-k}\mu_j\psi_{n-k-j}(x).$$

All terms in the last two summations on the right cancel out except for the $k = 0$ term to give

$$\frac{d^4\psi_n(x)}{dx^4} - \mu_0\psi_n(x) = \frac{2}{L}\sum_{k=0}^{n-1}\left(\frac{x}{L}\right)^k \frac{d^3\psi_{n-1-k}(x)}{dx^3} + \sum_{j=1}^{n}\mu_j\psi_{n-j}(x).$$

Substituting $k = j - 1$ in the first summation on the right-hand side we get the equation for the iterative perturbed tapered system (ITPS) which is

$$\frac{d^4\psi_n(x)}{dx^4} - \mu_0\psi_n(x) = \sum_{j=1}^{n}\left(\frac{2}{L}\left(\frac{x}{L}\right)^{j-1} \frac{d^3\psi_{n-j}(x)}{dx^3} + \mu_j\psi_{n-j}(x)\right). \tag{4.15}$$

4.4.1.1 Boundary Conditions of the ITPS

Equation (4.15) represents a set of fourth-order differential equations and should therefore accompany four boundary conditions for each of its realizations for a value of n. The boundary conditions for the nominal case when $n = 0$ are well known and match the boundary conditions of the rectangular beam. For the case of a fixed-free cantilever beam, the four boundary conditions are trivial and are given as

$$\begin{aligned} BC_1 &: \psi_{r,0}(0) = 0, \\ BC_2 &: \psi'_{r,0}(0) = 0, \\ BC_3 &: \psi''_{r,0}(L) = 0, \text{ and} \\ BC_4 &: \psi'''_{r,0}(L) = 0. \end{aligned} \tag{4.16}$$

Note that we have reintroduced the modal index r to illustrate that these boundary conditions are valid for the nominal component of each of the vibrational modes of the beam. Since the boundary conditions of the mode shapes of the tapered beam are completely catered by the nominal component's boundary conditions, the remaining perturbation components must not have any contribution to the total boundary conditions of the tapered beam and must, therefore, be set to zero [7]. For the nth perturbation component then, when $n \neq 0$, the boundary conditions can be given as

$$\begin{aligned} BC_1 &: \psi_{r,n}(0) = 0, \\ BC_2 &: \psi'_{r,n}(0) = 0, \\ BC_3 &: \psi''_{r,n}(L) = 0, \text{ and} \\ BC_4 &: \psi'''_{r,n}(L) = 0. \end{aligned} \tag{4.17}$$

Note that it is coincidental that these boundary conditions are identical to the nominal components' boundary conditions. If any of the boundary conditions of the nominal component is nonzero, such as in the case of a beam with a proofmass, the boundary conditions of the higher perturbation components would still be trivial as shown in Eq. (4.17).

4.4.2 Iterative Solution to the ITPS

We can use Eq. (4.15) to iteratively solve for $\psi_n(x)$ under the assumption that μ_n has been computed. The iterative solution starts with solving the equation for $n = 0$,

$$\frac{d^4\psi_0(x)}{dx^4} - \mu_0\psi_0(x) = 0. \tag{4.18}$$

The solution to Eq. (4.18) is the mode shape for a rectangular beam of which the mass normalized expression as derived in Sect. 2.6.3 is

$$\psi_0 = \frac{1}{\sqrt{Lm_0}}\left(\cosh\left(\frac{\lambda x}{L}\right) - \cos\left(\frac{\lambda x}{L}\right) - \sigma_r\left(\sinh\left(\frac{\lambda x}{L}\right) - \sin\left(\frac{\lambda x}{L}\right)\right)\right),$$

where $\lambda = L\mu^{\frac{1}{4}}$. The modal equation for $n = 1$ simplifies to

$$\frac{d^4\psi_1(x)}{dx^4} - \mu_0\psi_1(x) = \frac{2}{L}\frac{d^3\psi_0(x)}{dx^3} + \mu_1\psi_0(x),$$

which can be solved once we know μ_1. We will need to find a way to solve for μ_1 when $\psi_0(x)$ and μ_0 are known. Consequently, we will need a solution for each of μ_n assuming that we know $\{\mu_j\}_{j=0}^{n-1}$ and $\{\psi_j(x)\}_{j=0}^{n-1}$. We find this solution in Sect. 4.6. Moreover, we will need to solve Eq. (4.15) for $\psi_n(x)$ provided that we have solutions for $\{\psi_j(x)\}_{j=0}^{n-1}$ and $\{\mu_j\}_{j=0}^{n}$. We derive Green's function of Eq. (4.15) in Sect. 4.7 and use it to solve for $\psi_n(x)$ in Sect. 4.8. The iterative solution is illustrated in Fig. 4.7.

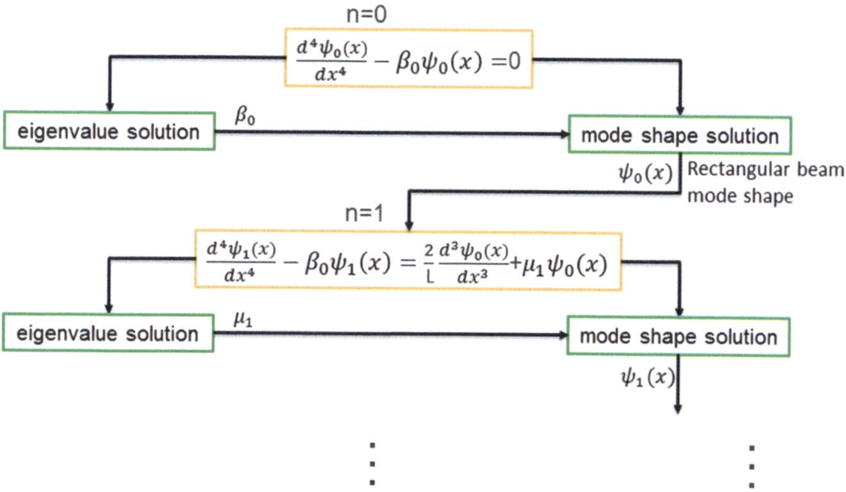

Fig. 4.7 Iterative solution structure to compute eigenvalues and mode shapes

4.5 Normalization Conditions

Now that we have the spatial modal equation expressed in a perturbation series expansion, we will proceed to derive normalization conditions for the tapered mode shapes that will be useful in removing free coefficients from the final solution. After obtaining the normalization conditions for the tapered mode shapes, we will determine what these normalization conditions mean for the individual perturbation components and derive some relationships between the different perturbation components.

4.5.1 Normalization Conditions for the Straight Cantilever Beam

The normalization conditions for a rectangular cantilever beam with a linear mass density of m and a flexural rigidity of YI are derived in [4] as

$$\int_0^L m\phi_r(x)\phi_s(x)dx = \delta_{rs}, \text{ and}$$

$$\int_0^L YI\phi_s(x)\frac{d^4\phi_r(x)}{dx^4}dx = \omega_r^2\delta_{rs}.$$

4.5.2 Normalization Conditions for the Tapered Beam Response

To arrive at the normalization conditions for the mode shapes of the tapered beam, we substitute mode-shape solutions for two distinct modes into Eq. (4.12). For mode index r, using the relationship $\beta_r = \frac{m_0\omega_r^2}{YI_0}$ we get

$$YI_0\frac{d^4\phi_r(x)}{dx^4} - m_0\omega_r^2\phi_r(x) + \alpha\left(-\frac{2}{L}YI_0\frac{d^3\phi_r(x)}{dx^3} - \frac{x}{L}YI_0\frac{d^4\phi_r(x)}{dx^4} + m_0\omega_r^2\frac{x}{L}\phi_r(x)\right) = 0.$$

Rearranging and using the relationships $YI(x) = YI_0(1-\alpha\frac{x}{L})$ and $m(x) = m_0(1-\alpha\frac{x}{L})$, we have

$$YI(x)\frac{d^4\phi_r(x)}{dx^4} - \frac{2}{L}\alpha YI_0\frac{d^3\phi_r(x)}{dx^3} - \omega_r^2 m(x)\phi_r(x) = 0. \tag{4.19}$$

Similarly, for a different eigensolution indexed by s we get

$$YI(x)\frac{d^4\phi_s(x)}{dx^4} - \frac{2}{L}\alpha YI_0\frac{d^3\phi_s(x)}{dx^3} - \omega_s^2 m(x)\phi_s(x) = 0. \tag{4.20}$$

Multiplying Eq. (4.19) with $\phi_s(x)$ and integrating over the beam length, we get

$$\int_0^L \phi_s(x)YI(x)\frac{d^4\phi_r(x)}{dx^4}dx - \frac{2}{L}\alpha YI_0\int_0^L \phi_s(x)\frac{d^3\phi_r(x)}{dx^3}dx = \omega_r^2\int_0^L \phi_s(x)m(x)\phi_r(x). \tag{4.21}$$

Using $\frac{dYI(x)}{dx} = -\frac{\alpha}{L}YI_0$ and simplifying over multiple applications of integration by parts, we get

$$\int_0^L \frac{d^2\phi_r(x)}{dx^2}\frac{d^2\phi_s(x)}{dx^2}YI(x)dx = \omega_r^2\int_0^L \phi_s(x)m(x)\phi_r(x). \tag{4.22}$$

Similarly multiplying Eq. (4.20) with $\phi_r(x)$ and following the same procedure, we get

$$\int_0^L \frac{d^2\phi_r(x)}{dx^2}\frac{d^2\phi_s(x)}{dx^2} YI(x)dx = \omega_s^2 \int_0^L \phi_s(x)m(x)\phi_r(x). \tag{4.23}$$

When r and s are distinct roots such that $\omega_r^2 \neq \omega_s^2$, we subtract Eq. (4.23) from Eq. (4.22) to get

$$\int_0^L \phi_r(x)m(x)\phi_s(x)dx = 0,\ r \neq s.$$

We establish the normalization criteria where the mass function weighted self-inner product of a mode shape is unity such that

$$\int_0^L \phi_r(x)^2 m(x)dx = 1,$$

which leads us to the following normalization conditions of a tapered beam.

The direct normalization condition is

$$\int_0^L \phi_r(x)m(x)\phi_s(x)dx = \delta_{rs}. \tag{4.24}$$

Using Eq. (4.22), we can obtain the companion form of the normalization condition as

$$\int_0^L \frac{d^2\phi_r(x)}{dx^2}\frac{d^2\phi_s(x)}{dx^2} YI(x)dx = \omega_r^2\delta_{rs}.$$

And using Eq. (4.21), we can obtain the alternate companion form of the normalization condition as

$$\int_0^L \phi_s(x)YI(x)\frac{d^4\phi_r(x)}{dx^4}dx - \frac{2}{L}\alpha YI_0 \int_0^L \phi_s(x)\frac{d^3\phi_r(x)}{dx^3}dx = \omega_r^2\delta_{rs}. \tag{4.25}$$

4.5.3 Normalization Conditions for Perturbation Components

We substitute perturbation expansion ($\phi_r(x) = \sum_{n=0}^{\infty} \psi_{r,n}(x)\alpha^n$) of the eigenfunctions from Eq. (4.13) in the normalization condition of Eq. (4.24) to get

$$\int_0^L \sum_{i=0}^{\infty} \psi_{r,i}(x)\alpha^i m(x) \sum_{j=0}^{\infty} \psi_{s,j}(x)\alpha^j dx = \delta_{rs}.$$

Substituting $m(x) = m_0(1 - \alpha\frac{x}{L})$ and redistributing the integral we get

$$m_0 \int_0^L \sum_{i=0}^{\infty}\sum_{j=0}^{\infty} \psi_{r,i}(x)\psi_{s,j}(x)\alpha^{i+j} dx$$
$$- \frac{m_0}{L} \int_0^L x \sum_{i=0}^{\infty}\sum_{j=0}^{\infty} \psi_{r,i}(x)\psi_{s,j}(x)\alpha^{i+j+1} dx = \delta_{rs}.$$

Comparing the coefficients of α^0, we get the normalization condition for the nominal perturbation component of a tapered beam as

$$m_0 \int_0^L \psi_{r,0}(x)\psi_{s,0}(x)dx = \delta_{rs}.$$

Comparing the coefficients of all terms with α^n and redistributing the summations, we get the normalization condition for the higher perturbation components of the tapered beam as

$$\int_0^L \psi_{r,0}(x)\psi_{s,n}(x)dx = \frac{1}{2l} \int_0^L x \sum_{i=0}^{n-1} \psi_{r,i}(x)\psi_{s,n-1-i}(x)dx$$
$$- \frac{1}{2} \int_0^L \sum_{i=1}^{n-1} \psi_{r,i}(x)\psi_{s,n-i}(x)dx. \tag{4.26}$$

4.5.3.1 Companion Form Normalization Conditions for Tapered Beam

We substitute the perturbation expansion of the eigenfunctions from Eq. (4.13) in the normalization condition of Eq. (4.25) to get

$$\int_0^L \sum_{j=0}^{\infty} \psi_{s,j}(x)\alpha^j YI(x) \frac{d^4 \sum_{j=0}^{\infty} \psi_{r,j}(x)\alpha^j}{dx^4} dx$$
$$- \frac{2}{L}\alpha Y I_0 \int_0^L \sum_{j=0}^{\infty} \psi_{s,j}(x)\alpha^j \frac{d^3 \sum_{j=0}^{\infty} \psi_{r,j}(x)\alpha^j}{dx^3} dx = \omega_r^2 \delta_{rs}.$$

Redistributing the summations and integrals, we get

$$\sum_{i=0}^{\infty}\sum_{j=0}^{\infty}\left(YI_0\alpha^{i+j} \int_0^L \psi_{s,i}(x)\frac{d^4\psi_{r,j}(x)}{dx^4}dx - YI_0\alpha^{i+j+1}\int_0^L \frac{x}{L}\psi_{s,i}(x)\frac{d^4\psi_{r,j}(x)}{dx^4}dx \right.$$
$$\left. -\frac{2}{L}\alpha^{i+j+1} Y I_0 \int_0^L \psi_{s,i}(x)\frac{d^3\psi_{r,j}(x)}{dx^3}dx \right) = \omega_r^2\delta_{rs}.$$

Comparing the coefficients of α^n on both sides of the equation we get the companion form normalization conditions of the perturbation components of the tapered beam shown in Eq. (4.27) as

$$\sum_{i=0}^{n} YI_0 \int_0^L \psi_{s,i}(x)\frac{d^4\psi_{r,n-i}(x)}{dx^4}dx - \frac{YI_0}{L}\sum_{i=0}^{n-1}\left(\int_0^L x\psi_{s,i}(x)\frac{d^4\psi_{r,n-i-1}(x)}{dx^4}dx \right.$$
$$\left. -2\int_0^L \psi_{s,i}(x)\frac{d^3\psi_{r,n-i-1}(x)}{dx^3}dx\right) = \omega_r^2\delta_{rs}. \tag{4.27}$$

4.6 Finding Eigenvalues of the Perturbed System

To solve for the resonance frequencies, we first evaluate the components of the eigenvalue series expansion in Eq. (4.14). Figure 4.8 helps establish the context to solve the eigenvalue components. The solution of the eigenvalues obtained here

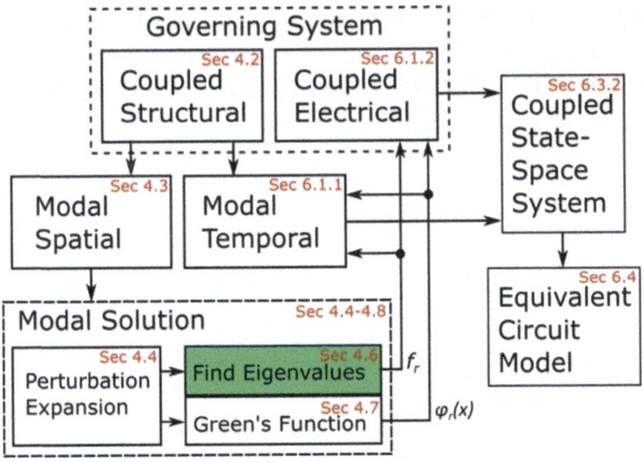

Fig. 4.8 Tapered PEH system formulation: finding the eigenvalues of the perturbed system

will interlace with the solution of the mode shapes via Green's function. We will first determine the eigenvalue of the nominal perturbation component which is the same as that of a rectangular cantilever beam. This is a well-known result [9] but is reproduced in Chap. 2 for completeness.

4.6.1 *Finding the 0th Perturbation Component of the Eigenvalue*

We substitute $n = 0$ in Eq. (4.15) and reintroduce the index r to get the nominal perturbation component of the rth mode:

$$\frac{d^4\psi_{r,0}(x)}{dx^4} - \mu_{r,0}\psi_{r,0}(x) = 0.$$

Substituting $\mu_{r,0} \rightarrow \left(\frac{\lambda_r}{L}\right)^4$ we get

$$\frac{\partial^4\psi_{r,0}(x)}{\partial x^4} - \left(\frac{\lambda_r}{L}\right)^4 \psi_{r,0}(x) = 0. \tag{4.28}$$

We have shown in Sect. 2.4 that the eigenvalues of Eq. (4.28) can be computed by numerically solving for the roots of the characteristic equation ($1 + \cos(\lambda_r)\cosh(\lambda_r) = 0$) of the rectangular cantilever system. The eigen-numbers for the first three modes are computed as $\lambda_1 = 1.8751$, $\lambda_2 = 4.6941$, and $\lambda_3 = 7.8548$, whereas the corresponding eigenvalue is computed as $\mu_{r,0} = \left(\frac{\lambda_r}{L}\right)^4$.

4.6.2 Eigenvalues of the Higher Perturbation Components

We drop the subscript r once again for clarity. The ITPS of equations from Eq. (4.15) is

$$\frac{d^4\psi_n(x)}{dx^4} - \mu_0\psi_n(x) = \sum_{j=1}^{n}\left(\frac{2}{L}\left(\frac{x}{L}\right)^{j-1}\frac{d^3\psi_{n-j}(x)}{dx^3} + \mu_j\psi_{n-j}(x)\right).$$

The homogeneous equation of the ITPS, which is also the perturbation equation for the nominal case at $n = 0$, is an eigenvalue problem of the form $(\mathcal{D} - \mu_0 I)\,\psi_n(x) = 0$ where $\mathcal{D} = \frac{d^4}{dx^4}$ and μ_0 its eigenvalue and $\psi_0(x)$, a solution to the rectangular beam mode-shape equation, its eigenvector. In addition, I is the multiplicative identity for the space of the operator $\mathcal{D}$. Let $\mathcal{L} = \mathcal{D} - \mu_0 I$, such that the ITPS can be written as $\mathcal{L}\psi_n(x) = F_n(x)$ where $F_n(x) = \sum_{j=1}^{n}\left(\frac{2}{L}\left(\frac{x}{L}\right)^{j-1}\frac{d^3\psi_{n-j}(x)}{dx^3} + \mu_j\psi_{n-j}(x)\right)$. We prove in Appendix A that the operator $\mathcal{L}$ is a Hermitian operator which implies that the null space of $\mathcal{L}$ and its image are orthogonal, that is, $\ker\mathcal{L} \perp \operatorname{Im}\mathcal{L}$. For the ITPS, this condition implies that $\psi_0(x) \perp F_n(x)$ which can be written as the orthogonality condition:

$$\langle F_n(x) \mid \psi_0(x)\rangle_x = 0, \tag{4.29}$$

where the inner product used is defined as

$$\langle u(x) \mid v(x)\rangle_x = \int_0^L u(x)v(x)dx. \tag{4.30}$$

We impose the condition of Eq. (4.29) on each instance of the ITPS for each value of n to derive each μ_n.

4.6.3 Evaluating Eigenvalue Perturbation Components Using $F_n(x)$ Orthogonality Property

Given that $F_n(x) \perp \psi_0(x)$, we can take the inner product of $\psi_0(x)$ with $F_n(x)$ and evaluate μ_n as the inner product evaluates to zero. We impose the orthogonality condition in Eq. (4.29) on the nth instance of the ITPS as

$$\begin{aligned}&\langle F_n(x) \mid \psi_0(x)\rangle_x\\&= \sum_{j=1}^{n}\left(\left\langle\frac{2}{L}\left(\frac{x}{L}\right)^{j-1}\frac{d^3\psi_{n-j}(x)}{dx^3} \mid \psi_0(x)\right\rangle_x + \mu_j\left\langle\psi_{n-j}(x) \mid \psi_0(x)\right\rangle_x\right) = 0.\end{aligned}$$

We separate μ_n from the summation and move it to the left-hand side to find its expression in terms of $\{\mu_j\}_{j=1}^{n-1}$ and $\{\psi_j(x)\}_{j=0}^{n-1}$.

$$\mu_n = \frac{-\sum_{j=1}^{n}\left\langle \frac{2}{L}\left(\frac{x}{L}\right)^{j-1}\frac{d^3\psi_{n-j}(x)}{dx^3} \mid \psi_0(x)\right\rangle_x - \sum_{j=1}^{n-1}\mu_j\left\langle\psi_{n-j}(x) \mid \psi_0(x)\right\rangle_x}{\langle\psi_0(x) \mid \psi_0(x)\rangle_x}.$$

The solution to the nominal case of a straight beam is mass-normalized such that the inner product of the mode shape with itself is

$$\langle\psi_0(x) \mid \psi_0(x)\rangle_x = \frac{1}{m_0}. \tag{4.31}$$

Reintroducing the mode number index r for completeness and using the normalization condition in Eq. (4.31), we get the expression in Eq. (4.32) below for the nth perturbation component of the rth eigenvalue.

$$\mu_{r,n} = -m_0\left(\sum_{j=1}^{n}\left\langle \frac{2}{L}\left(\frac{x}{L}\right)^{j-1}\frac{d^3\psi_{r,n-j}(x)}{dx^3} \mid \psi_{r,0}(x)\right\rangle_x + \sum_{j=1}^{n-1}\mu_{r,j}\left\langle\psi_{r,n-j}(x) \mid \psi_{r,0}(x)\right\rangle_x\right). \tag{4.32}$$

4.7 Green's Function of the Spatial Modal Equation

To solve for the nth perturbation component of the mode shape, $\psi_{r,n}(x)$, we have to solve the ITPS of equations. We can iteratively solve this system using the variation of parameters at each iteration. Alternatively, we can first find Green's function of the ITPS using the variation of parameters and then use it to solve for all successive components. Figure 4.9 shows the evaluation of Green's function in highlight to illustrate its place in the formulation of the tapered PEH system.

4.7.1 *Discovering Green's Function of the System*

Green's function is the solution to the system with a spatial impulse input concentrated at a point $x = \xi$. For the ITPS, Green's function is the solution to the boundary

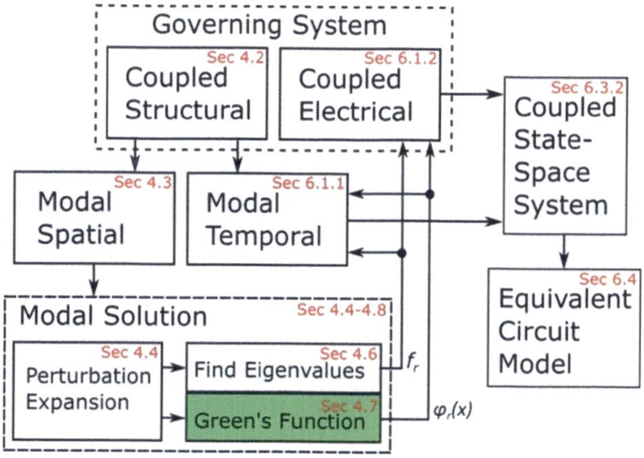

Fig. 4.9 Tapered PEH system formulation: evaluating Green's function of the perturbed system

value problem (BVP) with cantilever boundary conditions when the system is forced by an impulse at ξ, given as

$$\frac{d^4G(x,\xi)}{d\xi^4} - \mu_0 G(x,\xi) = \delta(x-\xi).$$

Since the Green function satisfies the homogeneous equation for $x \neq \xi$, it will be a linear combination of the homogenous solutions of the system. The linearly independent solutions to the homogeneous equation $\frac{d^4G(x,\xi)}{dx^4} - \mu_0 G(x,\xi) = 0$ identified in the solution to the nominal rectangular beam solution are $\sin\left(\frac{\lambda_r \xi}{L}\right)$, $\cos\left(\frac{\lambda_r \xi}{L}\right)$, $\sinh\left(\frac{\lambda_r \xi}{L}\right)$, and $\cosh\left(\frac{\lambda_r \xi}{L}\right)$. The solution for $G(x,\xi)$ will then be defined on either sides of $x = \xi$ as

$$G(x,\xi) = \begin{cases} c_1 \sin\left(\frac{\lambda_r \xi}{L}\right) + c_2 \cos\left(\frac{\lambda_r \xi}{L}\right) + c_3 \sinh\left(\frac{\lambda_r \xi}{L}\right) + c_4 \cosh\left(\frac{\lambda_r \xi}{L}\right) & 0 \le \xi < x \\ d_1 \sin\left(\frac{\lambda_r \xi}{L}\right) + d_2 \cos\left(\frac{\lambda_r \xi}{L}\right) + d_3 \sinh\left(\frac{\lambda_r \xi}{L}\right) + d_4 \cosh\left(\frac{\lambda_r \xi}{L}\right) & x < \xi \le L \end{cases}. \tag{4.33}$$

Here c_1 to c_4 and d_1 to d_4 are constants that are evaluated such that the boundary conditions of Green's function are satisfied.

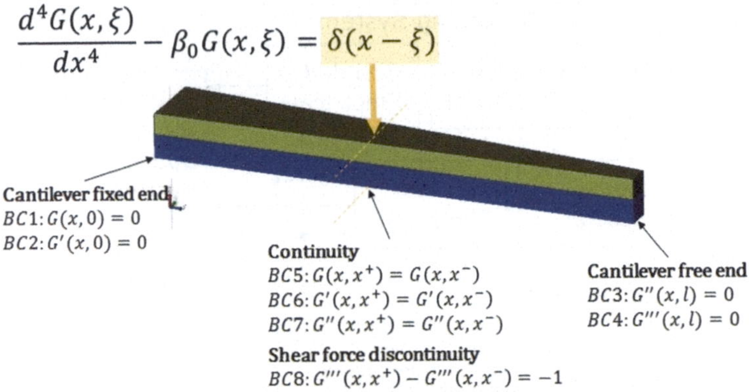

Fig. 4.10 Green's function boundary conditions. The derivatives shown are with respect to ξ

4.7.1.1 Green's Function Boundary Conditions

The eight boundary conditions satisfied by Green's function of the ITPS for a cantilever case are classified into three groups. These boundary conditions are illustrated in Fig. 4.10.

1. End Boundary Conditions Based on the Beam Type

The usual beam boundary conditions should hold at the ends. With the knowledge that Green's function is a two-part solution on either side of the applied impulse, we apply these end conditions to the part of Green's function evaluated there. BC_1 and BC_2 are boundary conditions of the clamped end, ensuring that the deflection and its slope are zero at the clamped end. BC_3 and BC_4 follow from the bending moment diagram in Fig. 4.11 which shows that the moment and its slope at the free end are both zero. These four boundary conditions are

$$
\begin{aligned}
&BC_1 : G(x,0) = 0,\\
&BC_2 : \left.\frac{\partial G(x,\xi)}{\partial \xi}\right|_{\xi=0} = 0,\\
&BC_3 : \left.\frac{\partial^2 G(x,\xi)}{\partial \xi^2}\right|_{\xi=L} = 0, \text{ and}\\
&BC_4 : \left.\frac{\partial^3 G(x,\xi)}{\partial \xi^3}\right|_{\xi=L} = 0.
\end{aligned}
$$

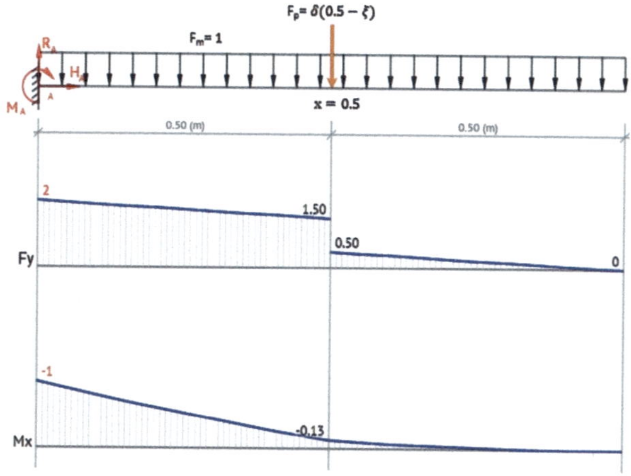

Fig. 4.11 Free body, shear force (Fy), and bending moment (Mx) diagrams of a cantilever beam. The beam is subjected to a distributed force of $F_m = 1Nm^{-1}$ representing the distributed mass and a point force of $F_p = 1N$ representing the impulse application at the halfway point

2. Continuity Conditions of Displacement, Slope, and Moment at $\xi = x$

The parts of Green's function on either side of the applied impulse must hold continuity and differentiability to have continuous slope and curvature at the point of the impulse. The bending moment diagram in Fig. 4.11 indicates that the moment at the position of the applied impulse also holds continuity thus imposing BC_3. These conditions are

$$BC_5 : G(x, x^+) = G(x, x^-),$$

$$BC_6 : \left.\frac{\partial G(x, \xi)}{\partial \xi}\right|_{\xi=x^+} = \left.\frac{\partial G(x, \xi)}{\partial \xi}\right|_{\xi=x^-}, \text{ and}$$

$$BC_7 : \left.\frac{\partial^2 G(x, \xi)}{\partial \xi^2}\right|_{\xi=x^+} = \left.\frac{\partial^2 G(x, \xi)}{\partial \xi^2}\right|_{\xi=x^-},$$

3. Shear Force Discontinuity of Magnitude One at $\xi = x$

Finally, the shear force that results from the applied impulse enforces a discontinuity in the shear force profile at the impulse as illustrated in the shear force diagram in Fig. 4.11. This is interpreted as a discontinuity in the third derivative of Green's function at $\xi = x$ and is given as

$$BC_8 : \left.\frac{\partial^3 G(x, \xi)}{\partial \xi^3}\right|_{\xi=x^+} - \left.\frac{\partial^3 G(x, \xi)}{\partial \xi^3}\right|_{\xi=x^-} = -1.$$

Applying the above boundary conditions, $BC_1 - BC_8$, to Green's function definition in Eq. (4.33) and solving for the unknown coefficients, we can find expressions of all the coefficients in terms of c_1 and c_2 as

$$
\begin{aligned}
c_3 &= -c_1, \\
c_4 &= -c_2, \\
d_1 &= c_1 + \frac{L^3}{2\lambda_r^3} \cos \frac{\lambda_r x}{L}, \\
d_2 &= c_2 - \frac{L^3}{2\lambda_r^3} \sin \frac{\lambda_r x}{L}, \\
d_3 &= -c_1 + \frac{L^3}{2\lambda_r^3} \cosh \frac{\lambda_r x}{L}, \\
d_4 &= -c_2 - \frac{L^3}{2\lambda_r^3} \sinh \frac{\lambda_r x}{L}.
\end{aligned}
$$

From BC_3 then, we have

$$
\begin{aligned}
c_2 =& - c_1 \left(\frac{\sin \lambda_r + \sinh \lambda_r}{\cos \lambda_r + \cosh \lambda_r} \right) + \frac{L^3}{2\lambda_r^3} \left(\frac{1}{\cos \lambda_r + \cosh \lambda_r} \right) \left(- \cos \frac{\lambda_r x}{L} \sin \lambda_r \right. \\
& \left. + \sin \frac{\lambda_r x}{L} \cos \lambda_r + \cosh \frac{\lambda_r x}{L} \sinh \lambda_r - \sinh \frac{\lambda_r x}{L} \cosh \lambda_r \right)
\end{aligned}
$$

or

$$
c_2 = -c_1 \left(\frac{\sin \lambda_r + \sinh \lambda_r}{\cos \lambda_r + \cosh \lambda_r} \right) + \frac{L^3}{2\lambda_r^3} \left(\frac{\sin \left(\frac{\lambda_r (x-L)}{L} \right) - \sinh \left(\frac{\lambda_r (x-L)}{L} \right)}{\cos \lambda_r + \cosh \lambda_r} \right).
$$

While ignoring BC_3 and using BC_4 instead we have

$$
\begin{aligned}
c_2 =& c_1 \left(\frac{\cos \lambda_r + \cosh \lambda_r}{\sin \lambda_r - \sinh \lambda_r} \right) + \frac{L^3}{2\lambda_r^3} \left(\frac{1}{\sin \lambda_r - \sinh \lambda_r} \right) \left(\cos \frac{\lambda_r x}{L} \cos \lambda_r \right. \\
& \left. + \sin \frac{\lambda_r x}{L} \sin \lambda_r - \cosh \frac{\lambda_r x}{L} \cosh \lambda_r + \sinh \frac{\lambda_r x}{L} \sinh \lambda_r \right)
\end{aligned}
$$

or

$$
c_2 = -c_1 \left(\frac{\cosh \lambda_r + \cos \lambda_r}{\sinh \lambda_r - \sin \lambda_r} \right) + \frac{L^3}{2\lambda_r^3} \left(\frac{\cosh \left(\frac{\lambda_r (x-L)}{L} \right) - \cos \left(\frac{\lambda_r (x-L)}{L} \right)}{\sinh \lambda_r - \sin \lambda_r} \right).
$$

Comparing the two answers we find that they are fundamentally different and this ambiguity is explained by a zero eigenvalue at resonance in Sect. 4.7.2. Hozhabrossadati had a similar observation about simply supported beams [10, 11].

4.7.2 *Proof That the System Has a Zero Eigenvalue*

The eigenvalue expansion view of Green's function explained in Appendix B shows that it can be expressed as

$$G(x,\xi)=\sum_{n=1}^{\infty}-\frac{v_n(x)v_n(\xi)}{\lambda_n\int_0^L v_n^2(\xi_1)d\xi_1}, \tag{4.34}$$

where $\{v_n(x)\}_{n=1}^{\infty}$ are the eigenvectors of the operator $\mathcal{L}=\frac{d^4}{dx^4}-\mu_0$ and λ_n its eigenvalues. We see from Eq. (4.34) that Green's function exists only when $\lambda_n\neq 0$. A zero eigenvalue will blow up Green's function and render it useless in finding the solution to the system. This will be the case when

$$\mathcal{L}u(x)=\frac{d^4u(x)}{d\xi^4}-\mu_0u(x)=-\lambda u(x)=0,$$

which will be true for a nonzero function $u(x)$ when a nontrivial solution exists for the homogenous problem:

$$\mathcal{L}u(x)=\frac{d^4u(x)}{d\xi^4}-\mu_0u(x)=0.$$

In the cantilever beam case, this is certainly true where the homogeneous problem is solved as the mode shape of a rectangular cantilever [4, 12] shown in Appendix 4. In the context of the ITPS, we identify this mode shape as the nominal perturbation component of the rth mode shape shown as

$$\begin{aligned}\psi_{r,0}(x)=&\frac{1}{\sqrt{Lm_0}}\left(\cosh\left(\frac{\lambda_r x}{L}\right)-\cos\left(\frac{\lambda_r x}{L}\right)\right.\\&\left.-\sigma_r\left(\sinh\left(\frac{\lambda_r x}{L}\right)-\sin\left(\frac{\lambda_r x}{L}\right)\right)\right).\end{aligned}$$

where $\sigma_r=\frac{\sinh(\lambda_r)-\sin(\lambda_r)}{\cosh(\lambda_r)+\cos(\lambda_r)}$ and $\left(\frac{\lambda_r}{L}\right)^4=\mu_{0,r}$.

4.7.3 Application of Fredholm Alternative Theorem

Fredholm's alternative theorem explained in Appendix B suggests that when a zero eigenvalue exists for a BVP, that is, it has a nontrivial null space, its inhomogeneous solution does not exist. The null space will then have to be deflated out of the system to solve for Green's function. The *modified Green's function*, $G_m(x, \xi)$, is the solution of the deflated problem:

$$\mathcal{L}G_m(x,\xi) = \frac{d^4 G_m(x,\xi)}{d\xi^4} - \mu_0 G_m(x,\xi) = \delta(x-\xi) + \kappa\psi_0(x) \times \psi_0(\xi).$$

Note that once again we have dropped the r subscript for the sake of clarity. Since $\psi_0(x)$ forms the null space of the system, it must be orthogonal to the image of the system $\mathcal{L}$, that is, $\ker\mathcal{L} \perp \mathrm{Im}\mathcal{L}$ and $\langle \delta(x-\xi) + \kappa\psi_0(x) \times \psi_0(\xi) \mid \psi_0(x)\rangle_\xi = 0$. This relationship allows us to find κ as

$$\int_0^L \psi_0(\xi)\left(\delta(x-\xi) + \kappa\psi_0(\xi)\psi_0(x)\right) d\xi = 0$$

$$\implies \psi_0(x) + \kappa\psi_0(x)\int_0^L \psi_0^2(\xi)d\xi = 0,$$

which gives us the Fredholm proportionality coefficient $\kappa = -\frac{1}{\int_0^L \psi_0^2(\xi)d\xi} = -m_0$. Hence, the modified BVP for the modified Green's function is written as

$$\frac{d^4 G_m(x,\xi)}{d\xi^4} - \mu_0 G_m(x,\xi) = \delta(x-\xi) - m_0\psi_0(x)\psi_0(\xi).$$

Using superposition and reintroducing the mode index r for completeness, the solution of this equation can be expressed as a sum of the impulse response, $G_{\delta,r}(x,\xi)$, and a particular solution $G_{0,r}(x,\xi)$ that represents the response to the null-space deflation of the system as

$$G_{m,r}(x,\xi) = G_{0,r}(x,\xi) + G_{\delta,r}(x,\xi).$$

Here $G_{\delta,r}(x,\xi)$ satisfies the equation

$$\frac{d^4 G_{\delta,r}(x,\xi)}{d\xi^4} - \mu_{r,0} G_{\delta,r}(x,\xi) = \delta(x-\xi), \tag{4.35}$$

and $G_{0,r}(x,\xi)$ satisfies the equation

$$\frac{d^4 G_{0,r}(x,\xi)}{d\xi^4} - \left(\frac{\lambda_r}{L}\right)^4 G_{0,r}(x,\xi) = -m_0\psi_{r,0}(x)\psi_{r,0}(\xi),$$

where $\mu_{r,0}$ has been replaced by $\left(\frac{\lambda_r}{L}\right)^4$.

4.7.4 Response to the Null-Space Deflation $G_{0,r}(x,\xi)$

Using the variation of parameters method we obtain the solution for $G_{0,r}(x,\xi)$ as

$$G_{0,r}(x,\xi) = \frac{L^2}{8\lambda_r^4}\sqrt{Lm_0}\psi_{r,0}(x)\left(3l\sqrt{Lm_0}\psi_{r,0}(\xi) + 2\lambda_r\xi\sigma_r\sqrt{Lm_0}\nu_{0,r}(\xi)\right), \tag{4.36}$$

where $\nu_{0,r}(x) = \frac{1}{\sqrt{Lm_0}}\left(\cosh\left(\frac{\lambda_r x}{L}\right) - \cos\left(\frac{\lambda_r x}{L}\right) - \frac{1}{\sigma_r}\left(\sinh\left(\frac{\lambda_r x}{L}\right) + \sin\left(\frac{\lambda_r x}{L}\right)\right)\right)$ and $\sigma_r = \frac{\sinh(\lambda_r)-\sin(\lambda_r)}{\cosh(\lambda_r)+\cos(\lambda_r)}$.

4.7.5 Impulse Response $G_{\delta,r}(x,\xi)$

The impulse response component of the modified Green's function is $G_{\delta,r}(x,\xi)$ which satisfies Eq. (4.35). $G_{\delta,r}(x,\xi)$ can be written as

$$G_{\delta,r}(x,\xi) = G_{\delta 1,r}(x,\xi)H(-\xi + x) + G_{\delta 2,r}(x,\xi)H(\xi - x),$$

where $G_{\delta 1,r}(x,\xi)$ is the component of $G_{\delta,r}(x,\xi)$ before the impulse and $G_{\delta 2,r}(x,\xi)$ is its component after the impulse. After expanding in terms of $G_{\delta 1,r}(x,\xi)$ and $G_{\delta 2,r}(x,\xi)$, the modified Green's function can be written as

$$G_{m,r}(x,\xi) = G_{0,r}(x,\xi) + G_{\delta,r}(x,\xi) = G_{0,r}(x,\xi) + \begin{cases} G_{\delta 1,r}(x,\xi) & 0 \le \xi < x \\ G_{\delta 2,r}(x,\xi) & x < \xi \le l \end{cases},$$

where

$$G_{\delta 1,r}(x,\xi) = c_1 \sin\left(\frac{\lambda_r\xi}{L}\right) + c_2\cos\left(\frac{\lambda_r\xi}{L}\right) + c_3\sinh\left(\frac{\lambda_r\xi}{L}\right) + c_4\cosh\left(\frac{\lambda_r\xi}{L}\right),$$

and

$$G_{\delta 2,r}(x,\xi) = d_1 \sin\left(\frac{\lambda_r\xi}{L}\right) + d_2\cos\left(\frac{\lambda_r\xi}{L}\right) + d_3\sinh\left(\frac{\lambda_r\xi}{L}\right) + d_4\cosh\left(\frac{\lambda_r\xi}{L}\right).$$

4.7.5.1 Boundary Conditions

We now apply the boundary conditions BC_1 through BC_8 to the modified Green's function. We would first like to find the contribution of $G_0(x, \xi)$ to the eight boundary conditions. For doing so we list the values of the relevant derivatives of $\psi_{r,0}(\xi)$ and $\nu_{0,r}(\xi)$ at the two endpoints:

At $\xi = 0$ we have

$$\begin{aligned}
\psi_{r,0}(0) &= 0, \\
\psi'_{r,0}(0) &= 0, \\
\nu_{0,r}(0) &= 0, \text{ and} \\
\nu'_{0,r}(0) &= \frac{2\lambda_r}{L\sigma_r\sqrt{Lm_0}}.
\end{aligned}$$

And at $\xi = L$ we have

$$\begin{aligned}
\psi''_{r,0}(L) &= \frac{2\lambda_r^2\,(1+\cos(\lambda_r)\cosh(\lambda_r))}{L^2\sqrt{Lm_0}\,(\cos(\lambda_r)+\cosh(\lambda_r))} = 0, \\
\psi'''_{r,0}(L) &= 0, \\
\nu'_{0,r}(L) &= \frac{2\lambda_r\,(1+\cos(\lambda_r)\cosh(\lambda_r))}{L\sqrt{Lm_0}\,(\sin(\lambda_r)-\sinh(\lambda_r))} = 0, \\
\nu''_{0,r}(L) &= 0, \text{ and} \\
\nu'''_{0,r}(L) &= \frac{2\lambda^3\sin(\lambda_r)\sinh(\lambda_r)}{L^3\sqrt{Lm_0}\,(\sin(\lambda_r)-\sinh(\lambda_r))}.
\end{aligned}$$

For the clamped end conditions, we get

$$\begin{aligned}
G_{0,r}(x,0) =& \frac{L^2}{8\lambda_r^4}\sqrt{Lm_0}\psi_{r,0}(x)\left(3l\sqrt{Lm_0}\psi_{r,0}(0)\right. \\
&\left. +2\lambda_r\cdot 0\cdot\sigma_r\sqrt{Lm_0}\nu_{0,r}(0).\right) \\
\Longrightarrow\ G_{0,r}(x,0) =& 0, \text{ and} \\
G'_{0,r}(x,0) =& \frac{L^2}{8\lambda_r^4}\sqrt{Lm_0}\psi_{r,0}(x)\left(3l\sqrt{Lm_0}\psi'_{r,0}(0)\right. \\
&\left. +2\lambda_r\sigma_r\sqrt{Lm_0}\left(0\cdot\nu'_{0,r}(0)+\nu_{0,r}(0)\right)\right) \\
\Longrightarrow\ G'_{0,r}(x,0) =& 0.
\end{aligned} \tag{4.37}$$

Moving on to the free end we look at the second and third derivatives of $G_{0,r}(x,\xi)$:

$$\begin{aligned}G''_{0,r}(x,L) =& \frac{L^2}{8\lambda_r^4}\sqrt{Lm_0}\psi_{r,0}(x)\left(3l\sqrt{Lm_0}\psi''_{r,0}(L)\right.\\ &\left.+2\lambda_r\sigma_r\sqrt{Lm_0}\left(l\nu''_{0,r}(L)+2\nu'_{0,r}(L)\right)\right)\\ \implies G''_{0,r}(x,L) =& 0.\end{aligned}$$

Finally, for BC_4 we need $\left.\frac{\partial^3 G_0(x,\xi)}{\partial\xi^3}\right|_{\xi=L}$ and we find that

$$\begin{aligned}\frac{\partial^3 G_0(x,\xi)}{\partial\xi^3} =& \frac{L^2}{8\lambda^4}\sqrt{Lm_0}\psi_{r,0}(x)\left(3l\sqrt{Lm_0}\psi'''_{r,0}(\xi)\right.\\ &\left.+2\lambda_r\sigma_r\sqrt{Lm_0}\left(\xi\nu'''_{0,r}(\xi)+3\nu''_{0,r}(\xi)\right)\right).\end{aligned}$$

We know that $\psi'''_{r,0}(L) = \psi''_{r,0}(L) = \nu''_{0,r}(L) = 0$ at resonance. We also know that $\nu'''_{0,r}(L) = \frac{2\lambda_r^3\sin(\lambda_r)\sinh(\lambda_r)}{L^3\sqrt{Lm_0}(\sin(\lambda_r)-\sinh(\lambda_r))}$ which gives the only nonzero contribution of $G_{0,r}(x,\xi)$ at resonance to the boundary conditions. Using $\sigma_r = \frac{\sinh(\lambda_r)-\sin(\lambda_r)}{\cosh(\lambda_r)+\cos(\lambda_r)}$, we get

$$\begin{aligned}\left.\frac{\partial^3 G_{0,r}(x,\xi)}{\partial\xi^3}\right|_{\xi=L} &= \frac{L^2}{8\lambda^4}\sqrt{Lm_0}\psi_{r,0}(x)\left(2\lambda_r\sigma_r\sqrt{Lm_0}\left(l\nu'''_{0,r}(L)\right)\right)\\ \left.\frac{\partial^3 G_{0,r}(x,\xi)}{\partial\xi^3}\right|_{\xi=L} &= \frac{1}{2}\sigma_r\sqrt{Lm_0}\psi_{r,0}(x)\left(\frac{\sin(\lambda_r)\sinh(\lambda_r)}{\sin(\lambda_r)-\sinh(\lambda_r)}\right)\\ \implies G'''_{0,r}(x,L) &= -\frac{1}{2}\sqrt{Lm_0}\psi_{r,0}(x)\left(\frac{\sin(\lambda_r)\sinh(\lambda_r)}{\cosh(\lambda_r)+\cos(\lambda_r)}\right).\end{aligned} \tag{4.38}$$

Using the contributions of $G_{0,r}(x,\xi)$ and its derivatives at the boundaries described in Eqs. (4.37)–(4.38), we get

$$\begin{aligned}&BC_1 \Rightarrow c_2+c_4=0,\\ &BC_2 \Rightarrow c_1+c_3=0,\\ &BC_3 \Rightarrow -d_1\sin(\lambda_r)-d_2\cos(\lambda_r)+d_3\sinh(\lambda_r)+d_4\cosh(\lambda_r)=0,\\ &BC_4 \Rightarrow -\frac{L^3}{2\lambda^3}\sqrt{Lm_0}\phi_r(x)\left(\frac{\sin(\lambda_r)\sinh(\lambda_r)}{\cosh(\lambda_r)+\cos(\lambda_r)}\right)\\ &\qquad -d_1\cos(\lambda_r)+d_2\sin(\lambda_r)+d_3\cosh(\lambda_r)+d_4\sinh(\lambda_r)=0,\\ &BC_5 \Rightarrow d_1\sin\frac{\lambda_r x}{L}+d_2\cos\frac{\lambda_r x}{L}+d_3\sinh\frac{\lambda_r x}{L}+d_4\cosh\frac{\lambda_r x}{L}\end{aligned}$$

$$
\begin{aligned}
&= c_1 \sin\frac{\lambda_r x}{L} + c_2 \cos\frac{\lambda_r x}{L} + c_3 \sinh\frac{\lambda_r x}{L} + c_4 \cosh\frac{\lambda_r x}{L}, \\
BC_6 \Rightarrow\ & d_1 \cos\frac{\lambda_r x}{L} - d_2 \sin\frac{\lambda_r x}{L} + d_3 \cosh\frac{\lambda_r x}{L} + d_4 \sinh\frac{\lambda_r x}{L} \\
&= c_1 \cos\frac{\lambda_r x}{L} - c_2 \sin\frac{\lambda_r x}{L} + c_3 \cosh\frac{\lambda_r x}{L} + c_4 \sinh\frac{\lambda_r x}{L}, \\
BC_7 \Rightarrow\ & -d_1 \sin\frac{\lambda_r x}{L} - d_2 \cos\frac{\lambda_r x}{L} + d_3 \sinh\frac{\lambda_r x}{L} + d_4 \cosh\frac{\lambda_r x}{L} \\
&= -c_1 \sin\frac{\lambda_r x}{L} - c_2 \cos\frac{\lambda_r x}{L} + c_3 \sinh\frac{\lambda_r x}{L} + c_4 \cosh\frac{\lambda_r x}{L}, \text{ and} \\
BC_8 \Rightarrow\ & -d_1 \cos\frac{\lambda_r x}{L} + d_2 \sin\frac{\lambda_r x}{L} + d_3 \cosh\frac{\lambda_r x}{L} + d_4 \sinh\frac{\lambda_r x}{L} \\
&= -c_1 \cos\frac{\lambda_r x}{L} + c_2 \sin\frac{\lambda_r x}{L} + c_3 \cosh\frac{\lambda_r x}{L} + c_4 \sinh\frac{\lambda_r x}{L} - \frac{L^3}{\lambda_r^3}.
\end{aligned}
$$

This system is underdetermined and hence we solve it for the coefficients in terms of c_1 using the first seven equations from BC_1 to BC_7 and substitute the results in BC_8 to test the solution. We will use the normalization conditions derived earlier to normalize c_1 out of the final mode-shape expression.

4.7.6 Solution for $G_8(x, \xi)$

The solution of the above system of boundary conditions gets quite cumbersome and it will take a few definitions to express the solution in a more concise manner. After selecting the analysis mode r and discovering the corresponding eigenvalues, we compute the modal constants as

$$
\begin{aligned}
\sigma_{0,r} &= \frac{\sinh\lambda_r - \sin\lambda_r}{\cosh\lambda_r + \cos\lambda_r}, \\
\sigma_{1,r} &= \frac{\sin\lambda_r \sinh\lambda_r}{\cosh\lambda_r \sin\lambda_r + \cos\lambda_r \sinh\lambda_r}, \\
\sigma_{2,r} &= \frac{\sin\lambda_r \sinh\lambda_r}{\cosh\lambda_r \sin\lambda_r - \cos\lambda_r \sinh\lambda_r}, \\
k_{1,r} &= \frac{L^3\sqrt{Lm_0}}{2\lambda_r^3}\,\frac{1}{\cosh\lambda_r + \cos\lambda_r}, \text{ and} \\
k_{2,r} &= k_1\left(\sinh\lambda_r + \sin\lambda_r\right).
\end{aligned}
\tag{4.39}
$$

We know that the solution to the beam equation is fundamentally a combination of trigonometric and hyperbolic functions. We, therefore, specify the trial solution set in the row vector $\mathbf{\Psi}_r(x)$ as

$$\mathbf{\Psi}_r(x) = \left[\cos\frac{\lambda_r x}{L}, \cosh\frac{\lambda_r x}{L}, \sin\frac{\lambda_r x}{L}, \sinh\frac{\lambda_r x}{L}\right].$$

We use the trial vector to define a list of fundamental beam shape functions that will help in describing Green's function as

$$\begin{aligned}
\psi_{r,0}(x) &= \frac{1}{\sqrt{Lm_0}}\left[1, -1, -\sigma_{0,r}, \sigma_{0,r}\right] \cdot \mathbf{\Psi}_r^T(x),\\
\nu_{0,r}(x) &= \frac{1}{\sqrt{Lm_0}}\left[1, -1, -\frac{1}{\sigma_{0,r}}, -\frac{1}{\sigma_{0,r}}\right] \cdot \mathbf{\Psi}_r^T(x),\\
\nu_{1,r}(x) &= \left[-\frac{1}{\sigma_{2,r}}, \frac{1}{\sigma_{2,r}}, 1, -1\right] \cdot \mathbf{\Psi}_r^T(x),\\
\nu_{2,r}(x) &= \left[-1, 1, \frac{1}{\sigma_{1,r}}, -\frac{1}{\sigma_{1,r}}\right] \cdot \mathbf{\Psi}_r^T(x),\\
\nu_{3,r}(x) &= [\sin\lambda_r, \sinh\lambda_r, -\cos\lambda_r, -\cosh\lambda_r] \cdot \mathbf{\Psi}_r^T(x),\\
\nu_{4,r}(x) &= [\sinh\lambda_r, -\sinh\lambda_r, \cos\lambda_r, \cosh\lambda_r] \cdot \mathbf{\Psi}_r^T(x), \text{ and}\\
\nu_{5,r}(x) &= [-\sin\lambda_r, \sin\lambda_r, \cos\lambda_r, \cosh\lambda_r] \cdot \mathbf{\Psi}_r^T(x).
\end{aligned} \tag{4.40}$$

Notice that $\psi_{r,0}(x)$ is the mode shape of the nominal perturbation component of the rth mode of the tapered beam. Also notice that we have already used $\psi_{r,0}(x)$ and $\nu_{0,r}(x)$ in the expression for $G_{0,r}(x,\xi)$ in Eq. (4.36). Since we would like to normalize out the free coefficient c_1 that will appear in the final expression for $G_{\delta,r}(x,\xi)$, we decide to collect all the terms in $G_{\delta,r}(x,\xi)$ that include c_1. Reordering the terms in $G_{\delta,r}(x,\xi)$ we can write it as

$$G_{\delta,r}(x,\xi) = c_1 G_{\delta a,r}(x,\xi) + G_{\delta b,r}(x,\xi).$$

We construct $G_{\delta a,r}(x,\xi)$ and $G_{\delta b,r}(x,\xi)$ by breaking each of $G_{\delta 1,r}(x,\xi)$ and $G_{\delta 2,r}(x,\xi)$ into their components that depend on c_1 and those that do not. $G_{\delta 1,r}(x,\xi)$ and $G_{\delta 2,r}(x,\xi)$ can then be expressed as

$$\begin{aligned}
G_{\delta 1,r}(x,\xi) &= c_1 G_{\delta 1a,r}(x,\xi) + G_{\delta 1b,r}(x,\xi)\\
G_{\delta 2,r}(x,\xi) &= c_1 G_{\delta 2a,r}(x,\xi) + G_{\delta 2b,r}(x,\xi).
\end{aligned}$$

$G_{\delta a,r}(x,\xi)$ and $G_{\delta b,r}(x,\xi)$ can then be expressed as

$$G_{\delta a,r}(x,\xi) = G_{\delta 1a,r}(x,\xi)H(-\xi + x) + G_{\delta 2a,r}(x,\xi)H(\xi - x)$$
$$G_{\delta b,r}(x,\xi) = G_{\delta 1b,r}(x,\xi)H(-\xi + x) + G_{\delta 2b,r}(x,\xi)H(\xi - x), \tag{4.41}$$

such that

$$G_{m,r}(x,\xi) = c_1 G_{\delta a,r}(x,\xi) + G_{\delta b,r}(x,\xi) + G_{0,r}(x,\xi). \tag{4.42}$$

$G_{\delta a,r}(x,\xi)$ and $G_{\delta b,r}(x,\xi)$ can be constructed given $G_{\delta 1a,r}(x,\xi)$, $G_{\delta 2a,r}(x,\xi)$, $G_{\delta 1b,r}(x,\xi)$, and $G_{\delta 2b,r}(x,\xi)$. To compute these functionals we define the following vectors of the elementary beam shape functions defined in terms of Eq. (4.40):

$$\boldsymbol{C}_{1,r}(x) = \left[1, -1, -\frac{\nu_{2,r}(x)}{\nu_{1,r}(x)}, \frac{\nu_{2,r}(x)}{\nu_{1,r}(x)}\right],$$
$$\boldsymbol{C}_{2,r}(x) = \left[0, 0, \frac{k_1\nu_{3,r}(x)\psi_{r,0}(x)}{\nu_{1,r}(x)}, -\frac{k_1\nu_{3,r}(x)\psi_{r,0}(x)}{\nu_{1,r}(x)}\right],$$
$$\boldsymbol{D}_{1,r}(x) = \left[1, -1, -\frac{\nu_{2,r}(x)}{\nu_{1,r}(x)}, \frac{\nu_{2,r}(x)}{\nu_{1,r}(x)}\right], \text{ and}$$
$$\boldsymbol{D}_{2,r}(x) = \left[\frac{k_2\sin\left(\frac{\lambda_r x}{L}\right)\psi_{r,0}(x)}{\nu_{1,r}(x)}, -\frac{k_2\sinh\left(\frac{\lambda_r x}{L}\right)\psi_{r,0}(x)}{\nu_{1,r}(x)}\right.$$
$$\left. -\frac{k_1\nu_{4,r}(x)\psi_{r,0}(x)}{\nu_{1,r}(x)}, \frac{k_1\nu_{5,r}(x)\psi_{r,0}(x)}{\nu_{1,r}(x)}\right]. \tag{4.43}$$

Using these vectors of beam shape functions we can compute $G_{\delta 1a,r}(x,\xi)$, $G_{\delta 1a,r}(x,\xi)$,$G_{\delta 1b,r}(x,\xi)$, and $G_{\delta 2b,r}(x,\xi)$ in the following way:

$$G_{\delta 1a,r}(x,\xi) = \mathbf{C}_{1,r}(x)\cdot\boldsymbol{\Psi}_r^T(\xi),$$
$$G_{\delta 1b,r}(x,\xi) = \mathbf{C}_{2,r}(x)\cdot\boldsymbol{\Psi}_r^T(\xi),$$
$$G_{\delta 2a,r}(x,\xi) = \mathbf{D}_{1,r}(x)\cdot\boldsymbol{\Psi}_r^T(\xi), \text{ and}$$
$$G_{\delta 2b,r}(x,\xi) = \mathbf{D}_{2,r}(x)\cdot\boldsymbol{\Psi}_r^T(\xi).$$

We can now find $G_{\delta a,r}(x,\xi)$ and $G_{\delta b,r}(x,\xi)$ using Eq. (4.41). Next, we will use the ITPS normalization conditions to find an expression for c_1 to finally evaluate the modified Green's function using Eq. (4.42).

4.8 Solution of the Perturbed Spatial Modal Equation

4.8.1 Nominal Component $\psi_{0,r}(x)$

The solution to the nominal component for $n = 0$ of the perturbed tapered beam system in Eq. (4.18) is a set of mass-normalized eigenfunctions $\psi_{r,0}(x)$ for the undamped cantilever beam and is shown to be [12]

$$\psi_{r,0}(x) = \frac{1}{\sqrt{Lm_0}}\left(\cosh\frac{\lambda_r x}{L} - \cos\frac{\lambda_r x}{L} - \sigma_{0,r}\left(\sinh\frac{\lambda_r x}{L} - \sin\frac{\lambda_r x}{L}\right)\right),$$

where $\sigma_r = \frac{\sinh\lambda_r - \sin\lambda_r}{\cosh\lambda_r + \cos\lambda_r}$ and $\mu_{r,0} = \left(\frac{\lambda_r}{L}\right)^4$.

4.8.2 Applying Green's Function to Solve Higher Perturbation Components

We find the general solution for the nth perturbed component by convolving Green's function with the forcing term on the right-hand side of the ITPS in Eq. (4.15) as

$$\psi_{r,n}(x) = \int_0^L G_{m,r}(x,\xi)F_{r,n}(\xi)d\xi. \tag{4.44}$$

We can determine $G_{m,r}(x,\xi)$ in terms of $G_{\delta a,r}(x,\xi)$, $G_{\delta b,r}(x,\xi)$, and the free parameter c_1. Substituting it in Eq. (4.44), we get

$$\psi_{r,n}(x) = c_1\int_0^L G_{\delta a,r}(x,\xi)F_{r,n}(s)ds + \int_0^L G_{\delta b,r}(x,\xi)F_{r,n}(s)ds.$$

4.8.2.1 Evaluating c_1 Using Normalization Conditions

Once we have computed $G_{\delta a,r}(x,\xi)$ and $G_{\delta b,r}(x,\xi)$, we can apply the perturbation components' normalization condition in Eq. (4.26) to recompute the value of c_1 iteratively as each perturbation component is evaluated. This normalization condition can be written using the inner product notation defined in Eq. (4.30) as

$$\begin{aligned}\langle\psi_{r,0}(x)\mid\psi_{r,n}(x)\rangle_x &= \frac{1}{2l}\sum_{i=0}^{n-1}\langle x\psi_{r,i}(x)\mid\psi_{r,n-1-i}(x)\rangle_x \\ &\quad -\frac{1}{2}\sum_{i=1}^{n-1}\langle\psi_{r,i}(x)\mid\psi_{r,n-i}(x)\rangle_x\,. \end{aligned}\tag{4.45}$$

Defining

$$\begin{aligned}\Phi_{r,n} =& \frac{1}{2l}\sum_{i=0}^{n-1}\langle x\psi_{r,i}(x)\mid\psi_{r,n-1-i}(x)\rangle_x \\ &-\frac{1}{2}\sum_{i=1}^{n-1}\langle\psi_{r,i}(x)\mid\psi_{r,n-i}(x)\rangle_x\,,\end{aligned}$$

simplifies Eq. (4.45) to

$$\langle\psi_{r,0}(x)\mid\psi_{r,n}(x)\rangle_x = \Phi_{r,n}.$$

With the knowledge that

$$\begin{aligned}\psi_{r,n}(x) =& \langle G_{m,r}(x,\xi)\mid F_{r,n}(\xi)\rangle_\xi = c_1\langle G_{\delta a,r}(x,\xi)\mid F_{r,n}(\xi)\rangle_\xi \\ &+\langle G_{\delta b,r}(x,\xi)+G_{0,r}(x,\xi)\mid F_{r,n}(\xi)\rangle_\xi\,,\end{aligned}$$

we expand Eq. (4.45) to get

$$\begin{aligned}&c_1\left\langle\psi_{r,0}(x)\mid\langle G_{\delta a,r}(x,\xi)\mid F_{r,n}(\xi)\rangle_\xi\right\rangle_x \\ &\quad+\left\langle\psi_{r,0}(x)\mid\langle G_{\delta b,r}(x,\xi)+G_{0,r}(x,\xi)\mid F_{r,n}(\xi)\rangle_\xi\right\rangle_x = \Phi_{r,n}.\end{aligned}$$

We can then evaluate c_1 as

$$c_1 = \frac{\Phi_{r,n} - \left\langle\psi_{r,0}(x)\mid\langle G_{\delta b,r}(x,\xi)+G_{0,r}(x,\xi)\mid F_{r,n}(\xi)\rangle_\xi\right\rangle_x}{\left\langle\psi_{r,0}(x)\mid\langle G_{\delta a,r}(x,\xi)\mid F_{r,n}(\xi)\rangle_\xi\right\rangle_x}$$

Plugging c_1 back in the expression for $\psi_{r,n}(x)$ allows us to normalize the nth perturbation component of the mode shape appropriately such that the final mode shape is mass-normalized. Once the perturbed components are evaluated, the net mode shape is evaluated as

$$\phi_r(x) = \sum_{n=0}^{\infty} \psi_{r,n}(x)\alpha^n,$$

and the eigenvalues can be evaluated as

$$\beta_r = \beta_{r,0} + \sum_{n=1}^{\infty} \mu_{r,n}\alpha^n.$$

4.9 Tapered Beam Modal Analysis Algorithm

The modal analysis of a tapered beam is summed up in Algorithm 4.1. The procedure works on the specified mode number r, at the given spatial sample locations x, using N_p perturbation expansion terms. The algorithm first finds the nominal eigenvalue and mode shape for the rth mode and then sets up the perturbation problem to derive Green's function of the nominal system to compute each perturbation component's mode-shape contribution. Once the mode-shape components are normalized, they can be synthesized to form the total tapered beam mode shape. At the end, the mode eigenvalue is computed leading to an estimate of the resonance frequency.

The modal analysis of a tapered cantilever beam is summarized in Algorithm 4.1 and Fig. 4.12. The procedure works on the specified mode number r, at the given spatial samples $x = \{x_i\}_{i=1}^{N_x} : x_i \in (0, L)$ where N_x is the number of spatial samples, using N_p perturbation expansion terms. The algorithm first finds the nominal eigenvalue and mode shape for the rth mode and then sets up the perturbation problem to derive the Green function of the nominal system to compute the mode-shape contribution of each perturbation component. Once the mode-shape components are normalized, they can be put together to form the total tapered beam mode shape. In the end, the mode eigenvalue is computed, leading to an estimate of the resonance frequency.

Some of the advantages of the proposed algorithm compared to some of the solutions found in the literature are listed below:

1. As illustrated in Sect. 5.6, the proposed solution is capable of evaluating the mode shapes and natural frequencies reliably for the whole range of taper factors ($0 \leq \alpha < 1$). The solution in [13] fails for $\alpha > 0.5$ as the beam shape approaches a triangle, and the solution in [14] gives unsatisfactory results for $\alpha < 0.25$ as the beam shape approaches a rectangle. Our method is also equally capable of

Algorithm 4.1 Modal analysis of a tapered cantilever using perturbation expansion

procedure MODALANALYSISOFTAPEREDCANTILEVER($beam,\ prop,\ r,\ N_p,\ x$)
$beam$: Beam geometry ($l, b_0, b_L, h_i \forall i \in (1, number\ of\ layers)$), $prop$: Material properties, r: Mode number, N_p: Perturbation order, $x = \{x_i\}_{i=1}^{N_x},\ where\ x_i \in (0, l)$

Solve for nominal perturbation component, $\{\psi_{r,0}(x),\ \mu_{r,0}\}$
Numerically solve $\cos\lambda_r \cosh\lambda_r + 1 = 0$ ▷ $\lambda_1 = 1.8751$
$\psi_{r,0}(x) = \frac{1}{\sqrt{Lm_0}}\left(\cosh\frac{\lambda_r x}{L} - \cos\frac{\lambda_r x}{L} - \frac{\sinh\lambda_r - \sin\lambda_r}{\cosh\lambda_r + \cos\lambda_r}\left(\sinh\frac{\lambda_r x}{L} - \sin\frac{\lambda_r x}{L}\right)\right)$
$\mu_{r,0} = \left(\frac{\lambda_r}{L}\right)^4$

Setup
Compute α ▷ Taper factor: $\alpha = \frac{b_0 - b_L}{b_0}$
Homogenize layers: Compute YI_0, Y_0, I_0, m_0
Define $\sigma_0, \sigma_1, \sigma_2, k_1, k_2$ ▷ Use Eq. (4.39)
Define $\Psi_{trials,r}(x) = \left[\cos\frac{\lambda_r x}{L}, \cosh\frac{\lambda_r x}{L}, \sin\frac{\lambda_r x}{L}, \sinh\frac{\lambda_r x}{L}\right]$
Define $\boldsymbol{C}_{1,r}(x), \boldsymbol{C}_{2,r}(x), \boldsymbol{D}_{1,r}(x), \boldsymbol{D}_{2,r}(x)$ ▷ Use Eq. (4.43)

Evaluate Green's Function
Null-space deflation response , $G_{0,r}(x, \xi)$
$G_{0,r}(x,\xi) = \frac{L^2}{8\lambda_r^4}\sqrt{Lm_0}\psi_{0,r}(x)\left(3l\sqrt{Lm_0}\psi_{0,r}(\xi) + 2\lambda_r\xi\sigma_0\sqrt{Lm_0}\nu_{0,r}(\xi)\right)$
Impulse response , $G_{\delta,r}(x,\xi)$
$G_{\delta 1a,r}(x,\xi) = \boldsymbol{C}_{1,r}(x)\Psi_{trials,r}^T(\xi),\ G_{\delta 1b,r}(x,\xi) = \boldsymbol{C}_{2,r}(x)\Psi_{trials,r}^T(\xi)$
$G_{\delta 2a,r}(x,\xi) = \boldsymbol{D}_{1,r}(x)\Psi_{trials,r}^T(\xi),\ G_{\delta 2b,r}(x,\xi) = \boldsymbol{D}_{2,r}(x)\Psi_{trials,r}^T(\xi)$
$G_{\delta a,r}(x,\xi) = G_{\delta 1a,r}(x,\xi)H(-\xi + x) + G_{\delta 2a,r}(x,\xi)H(\xi - x)$
$G_{\delta b,r}(x,\xi) = G_{\delta 1b,r}(x,\xi)H(-\xi + x) + G_{\delta 2b,r}(x,\xi)H(\xi - x)$

Solve for higher perturbation components, $\{\psi_{r,n}(x),\ \mu_{r,n}\}_{n=1}^{N_p}$
for $n \leftarrow 1, N_p$ **do**
$D_{r,n,j}(x) = \frac{2}{L}\left(\frac{x}{L}\right)^{j-1}\psi_{r,n-j}'''(x)$
Eigenvalue perturbation component estimate, $\mu_{r,n}$
$\mu_{r,n} = -m_0\left(\sum_{j=1}^{n}\langle D_{r,n,j}(x) \mid \psi_{r,0}(x)\rangle_x + \sum_{j=1}^{n-1}\mu_{r,j}\langle\psi_{r,n-j}(x) \mid \psi_{r,0}(x)\rangle_x\right)$
Forcing function estimate, $F_{r,n}(x)$
$F_{r,n}(x) = \sum_{j=1}^{n}\left(D_{r,n,j}(x) + \mu_{r,j}\psi_{r,n-j}(x)\right)$
Normalization coefficient from BCs, c_1
$\Phi_{r,n} = \frac{1}{2l}\sum_{i=0}^{n-1}\langle x\psi_{r,i}(x) \mid \psi_{s,n-1-i}(x)\rangle_x - \frac{1}{2}\sum_{i=1}^{n-1}\langle\psi_{r,i}(x) \mid \psi_{s,n-i}(x)\rangle_x$
$c_1 = \frac{\Phi_{r,n} - \left\langle\psi_{r,0}(x) \mid \langle G_{\delta b,r}(x,\xi) + G_{0,r}(x,\xi) \mid F_{r,n}(\xi)\rangle_\xi\right\rangle_x}{\left\langle\psi_{r,0}(x) \mid \langle G_{\delta a,r}(x,\xi) \mid F_{r,n}(\xi)\rangle_\xi\right\rangle_x}$
Mode-shape perturbation component estimate, $\psi_{r,n}(x)$
$\psi_{r,n}(x) = \langle c_1 G_{\delta a,r}(x,\xi) + G_{\delta b,r}(x,\xi) + G_{0,r}(x,\xi) \mid F_{r,n}(\xi)\rangle_\xi$
end for
Return $\hat{\phi}_{r,N_p}(x) = \sum_{n=0}^{N_p}\psi_{r,n}(x)\alpha^n$ ▷ Mode shape estimate
$\hat{\beta}_{r,N_p} = \mu_{r,0} + \sum_{n=1}^{N_p}\mu_{r,n}\alpha^n$ ▷ Eigenvalue estimate

Return $\hat{\omega}_{r,N_p} = \sqrt{\frac{YI_0\hat{\beta}_{r,N_p}}{m_0}}$ ▷ Resonance frequency estimate
end procedure

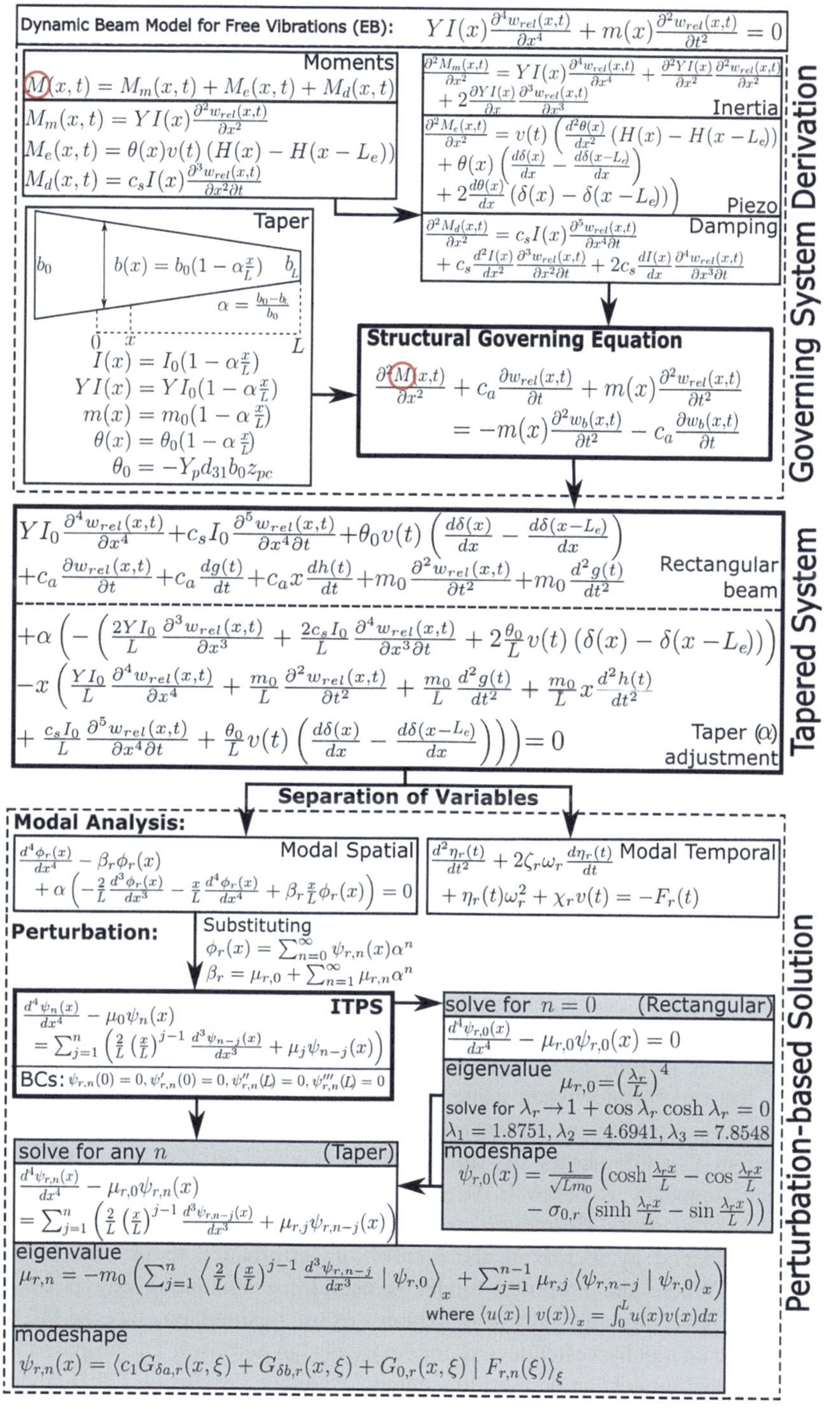

Fig. 4.12 System formulation equation map

evaluating higher resonance modes for as long as enough number of samples are chosen along the beam length.

2. Unlike [14] where the resonance frequency computation starts with an arbitrary guess before it is iteratively updated and refined, our method computes the resonance frequency methodically starting from the frequency solution of a rectangular beam before updating it using rules established by the perturbation system.
3. In contrast to the approximate series solutions found in literature where the power series is either of the system eigenvalue [15] or of the x-coordinate [14, 16], we have used perturbation expansion to set up a power series of the taper factor α. Since the taper factor is always less than one ($\alpha < 1$) with the limiting cases of $\alpha = 0$ for a rectangular beam and $\alpha = 1$ for a triangular beam, the power series is inherently convergent, especially since the adjustments to the mode shapes and eigenvalues of the rectangular beam solution become progressively smaller in the higher-order terms.
4. Since the proposed solution is a perturbation expansion around the rectangular beam solution, the first expansion term holds a physical significance that can be very useful for beam designers in that it gives them a direct reference to measure the impact of the taper on the design.

4.10 Conclusion

We have developed a novel semi-analytical model of and its solution to the vibrational dynamics of a multilayered tapered prismatic microcantilever. We have demonstrated that this methodology can be used to perform modal analyses on arbitrarily tapered trapezoidal and triangular cantilevers. We hold off on presenting the validation of the proposed semi-analytical tapered beam model until Chap. 5 where we further elaborate on the methodology that we used to arrive at this model before presenting its performance validation. The computational time comparisons presented in Chap. 5 show that the proposed semi-analytical solution enables the modal analysis to run $\times 300$ times faster when using 20 perturbation terms and up to $\times 700$ times faster when using 4 perturbation terms when compared with a computationally efficient commercial FEM solution. The error analysis, also presented in Chap. 5, shows that the model prediction error of the first mode resonance frequency is below 0.1% for 20 perturbation components and below 10% for 4 components. Estimation errors increase with increasing taper. However, they are managed by increasing the number of perturbation terms used for the estimation. Where the proposed model is useful for simulating any prismatic tapered cantilever structure, this work explores its use in simulating tapered MEMS piezoelectric energy harvester devices for device design purposes. The proposed fast semi-analytical model can be used to simulate candidate designs of TPEH devices in a design optimization loop. The algorithm structure suggests that an adaptive simulation strategy can be set up in which fewer perturbation components are used for structures with lower taper, and more perturbation components are used for higher-taper structures.

References

1. William Weaver Jr, Stephen P Timoshenko, and Donovan Harold Young. *Vibration problems in engineering*. John Wiley & Sons, 1990.
2. S Timoshenko and DH Young. "Vibration Problems in Engineering, (1955)". In: *Van Nostrana* (1968).
3. AH Meitzler et al. *IEEE standard on piezoelectricity*. 1988.
4. Alper Erturk and Daniel J Inman. "A distributed parameter electromechanical model for cantilevered piezoelectric energy harvesters". In: *Journal of vibration and acoustics* 130.4 (2008), p. 041002.
5. Dumitru I Caruntu. "Eigenvalue singular problem of factorized fourth-order self-adjoint differential equations". In: *Applied Mathematics and Computation* 224 (2013), pp. 603–610.
6. John K Hunter. "Asymptotic analysis and singular perturbation theory". In: *Department of Mathematics, University of California at Davis* (2004).
7. Steven A Orszag and CM Bender. "Advanced mathematical methods for scientists and engineers". In: *Mac Graw Hill* (1978).
8. AH Nayfeh. *Perturbation Methods*. John Wiley & Sons Inc., New York, 1973.
9. Alper Erturk and Daniel J Inman. "On mechanical modeling of cantilevered piezoelectric vibration energy harvesters". In: *Journal of Intelligent Material Systems and Structures* (2008).
10. Seyed Mojtaba Hozhabrossadati et al. "Green's function for uniform Euler-Bernoulli beams at resonant condition: Introduction of Fredholm alternative theorem". In: *Applied Mathematical Modelling* 39.12 (2015), pp. 3366–3379.
11. Seyed Mojtaba Hozhabrossadati, Ahmad Aftabi Sani, and Masood Mofid. "Resonant Green's function for Euler-Bernoulli beams by means of the Fred-holm alternative theorem". In: *Journal of Applied Mathematics and Computational Mechanics* 12.3 (2013), pp. 55–67.
12. TK Caughey and MEJ O'kelly. "Classical normal modes in damped linear dynamic systems". In: *Journal of Applied Mechanics* 32.3 (1965), pp. 583–588.
13. Serge Abrate. "Vibration of non-uniform rods and beams". In: *Journal of sound and vibration* 185.4 (1995), pp. 703–716.
14. S Naguleswaran. "Vibration of an Euler-Bernoulli beam of constant depth and with linearly varying breadth". In: *Journal of sound and vibration* 153.3 (1992), pp. 509–522.
15. Shu-qi Guo and Shao-pu Yang. "Transverse vibrations of arbitrary non-uniform beams." In: *Applied Mathematics & Mechanics* 35.5 (2014).
16. Jung-Chang Hsu, Hsin-Yi Lai, and Cha'o-Kuang Chen. "Free vibration of non-uniform Euler–Bernoulli beams with general elastically end constraints using Adomian modified decomposition method". In: *Journal of Sound and Vibration* 318.4-5 (2008), pp. 965–981.

Chapter 5
Symbolic Model Order Reduction Using Perturbation Analysis

In this chapter, a novel symbolic modeling methodology is presented for MEMS devices having beams with tapered geometries. Such beams are very common in piezoelectric energy harvesters. While a closed-form solution for the device model may be possible in cases without taper, it becomes increasingly difficult to obtain for tapered devices. Symbolic perturbation analysis is employed where the tapering slope is used as a symbolic perturbation parameter around the nominal case of no taper. It is shown that the symbolic perturbation expansion as a power series in the taper slope has an appealing iterative structure where each term of the expansion can be symbolically generated from the previous terms via symbolic Green's functions. To facilitate such symbolic power series expansion, a specialized library of Mathematica modules that comprise the full symbolic modeling flow from device setup to the term-by-term derivations of Green's functions and the modal solutions of the tapered beam has been developed. Symbolic model evaluations are performed using MATLAB scripts that enable fast design space exploration and the development of equivalent lumped models for efficient device simulations. The symbolic model has tunable accuracy and can result in orders of magnitude speedup compared to commercial FEM solvers.

5.1 Introduction

Tapered shapes are widely used in engineering due to their superior structural properties [1]. Microelectromechanical devices have profited from these properties to enhance the reliability of such important components as interdigitated combs, springs, and multilayered Euler–Bernoulli beams. In particular, piezoelectric vibrational energy harvesters have widely employed tapered beams. Their successful design is heavily dependent on an accurate and efficient model of the tapered geometry. Due to the complexity added by the shape taper, these MEMS devices

W. U. Syed, I. (Abe) M. Elfadel, *Tapered Beams in MEMS*,
https://doi.org/10.1007/978-3-031-66391-8_5

are modeled using the finite element method (FEM) of such commercial tools as CoventorWare and MEMS+ from Coventor [2]. Our previous work [3] described a lumped semi-analytical model of a tapered beam. The model was intended to accelerate the design of multilayer, tapered piezoelectric energy harvesters (TPEHs). In this chapter, we describe the symbolic model construction and solution methodology that are at the foundation of the lumped, semi-analytical TPEH model. Their underlying mathematics is that of symbolic perturbation analysis with the modeling methodology consisting of several *symbolic* computational modules organized around specialized application programming interfaces (APIs) coded in the symbolic language of Mathematica. In addition, *numerical* modules scripted in MATLAB are used to evaluate the symbolic models, explore the TPEH design space, and select specific design candidates.

Symbolic methods in modeling have a long history that goes back to the dawn of computer-aided design (CAD). The specific area that has benefited most from these methods is that of analog circuit design, where symbolic modeling tools such as ISAAC [4] have found significant use in relation to the derivation of transfer functions for analog linear filters in both the Laplace domain (op-amp circuits) and the z domain (switched-capacitance circuits). The early history of symbolic modeling in analog circuit design is summarized in [5], and some of its advanced methods are surveyed in [6]. This tradition of symbolic modeling in analog circuits continues to the present day, with ever more complex analog circuits being designed using insights gained from symbolic models [7–9]. This well-established CAD tradition in symbolic modeling has two main limitations. The first is that it is focused on designs that can be described with lumped-parameter models. The second is the fragmentation of the symbolic analysis tools due to the absence of a unified symbolic modeling language. One objective of this chapter is to address these two limitations in the context of MEMS modeling by (a) attacking the symbolic modeling problem of a MEMS component whose fundamental description uses a complex distributed model, namely, a fourth-order partial differential equation, and (b) adopting the commonly used symbolic language of Mathematica to implement the modeling methodology.

In summary, our main contributions presented in this chapter are as follows:

1. We develop a symbolic methodology for the modeling of tapered Euler–Bernoulli beams using symbolic perturbation analysis with respect to the tapering coefficient of the beam.
2. We fully expose the algorithmic details of symbolically computing the terms of the power series expansion using a novel iterative approach for the term-by-term derivation of Green's functions.
3. We use the symbolic language of Mathematica to implement the distributed symbolic model.
4. Finally, we numerically evaluate the symbolic perturbation model using MATLAB scripts and show that it can achieve orders of magnitude speedup with respect to commercial FEM tools.

The exposition of this chapter emphasizes the underlying mathematics of the symbolic perturbation expansion and presents the algorithmic counterpart of our recent publication [10] which focused on the symbolic software implementation of the proposed modeling methodology. Whereas the math presented in this chapter has a similar look and feel as to that presented in Chap. 4, it serves the purpose of illustrating the symbolic framework of the methodology used to derive these equations.

The remainder of this chapter is organized as follows. In Sect. 5.2, we provide background information on the traditional Euler–Bernoulli beam that will prove useful for the initialization step of the symbolic model. Section 5.3 presents the fundamental idea of the symbolic perturbation analysis of the tapered beam together with the Mathematica code used to implement it. Section 5.4 is devoted to the core iterative symbolic algorithm used to provide a full modal solution of the tapered beam. The MATLAB modules used for numerical model evaluation are presented in Sect. 5.5. Numerical results illustrating the order-of-magnitude gain in performance with respect to commercial FEM tools are summarized in Sect. 5.6. The chapter is concluded in Sect. 5.7.

5.2 Background

Figure 5.1 shows the parameterized device geometry of a simple cantilever TPEH with a substrate and a piezoelectric layer, having length l, thickness h, base width b_0, and tip width b_l. The taper coefficient α is calculated as

$$\alpha = \frac{b_0 - b_l}{b_0}.$$

Recall from Chap. 2 that the governing equation of motion for undamped free vibrations of a uniform beam ($\alpha = 0$) is the dynamic Euler–Bernoulli beam equation

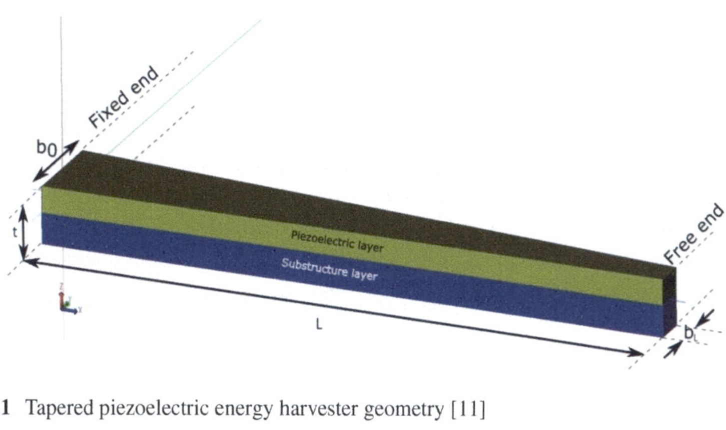

Fig. 5.1 Tapered piezoelectric energy harvester geometry [11]

given by Erturk and Inman [12]

$$YI\frac{\partial^4 w_{rel}(x,t)}{\partial x^4} + m\frac{\partial^2 w_{rel}(x,t)}{\partial t^2} = 0, \tag{5.1}$$

and the clamped-free boundary conditions of the cantilever beam without any proof mass are defined at $x = 0$ and $x = l$ as

$$BC_1 : w_{rel}(0,t) = 0,$$

$$BC_2 : \left.\frac{\partial w_{rel}(x,t)}{\partial x}\right|_{x=0} = 0,$$

$$BC_3 : \left.\frac{\partial^2 w_{rel}(x,t)}{\partial x^2}\right|_{x=l} = 0, \text{ and}$$

$$BC_4 : \left.\frac{\partial^3 w_{rel}(x,t)}{\partial x^3}\right|_{x=l} = 0.$$

Using the separation of variables,

$$w_{rel}(x,t) = \sum_{r=1}^{\infty} \eta_r(t)\phi_r(x),$$

and substituting in Eq. (5.1) to obtain, for each mode r, the equation

$$YI\eta_r(t)\frac{\partial^4 \phi_r(x)}{\partial x^4} + m\phi_r(x)\frac{\partial^2 \eta_r(t)}{\partial t^2} = 0,$$

which, along with $\frac{\partial^2 \eta_r(t)}{\partial t^2} = -\omega_r^2 \eta_r(t)$ and for a harmonic load at frequency ω_r, gives a differential equation in the spatial domain that describes the mode shape $\phi_r(x)$ as

$$\frac{d^4 \phi_r(x)}{dx^4} - \beta_r \phi_r(x) = 0, \tag{5.2}$$

where $\beta_r = \frac{m\omega_r^2}{IY}$. Note that the pair (β_r, ϕ_r) constitutes an (eigenvalue, eigenfunction) pair for the beam's spatial behavior.

The starting point of our symbolic modeling methodology is the fundamental observation that for a tapered Euler–Bernoulli beam with the taper coefficient α, the differential equation of the rth mode shape becomes

$$\frac{d^4 \phi_r(x)}{dx^4} - \beta_r \phi_r(x)$$

$$+\alpha\left(-\frac{2}{l}\frac{d^3\phi_r(x)}{dx^3}-\frac{x}{l}\frac{d^4\phi_r(x)}{dx^4}+\beta_r\frac{x}{l}\phi_r(x)\right)=0, \tag{5.3}$$

under a slightly modified definition of the eigenvalue β_r. Clearly, Eq. (5.3) reduces to Eq. (5.2) when $\alpha = 0$.

Remark Note that in the presence of taper α, physical attributes, such as mass density m and moment of inertia I, become dependent on the spatial variable x along the length of the beam, according to the relations

$$m(x) = m_0\left(1-\alpha\frac{x}{l}\right), \text{ and}$$

$$I(x) = I_0\left(1-\alpha\frac{x}{l}\right).$$

The governing equations of the tapered beam take into account this dependence.

5.3 Symbolic Perturbation Expansion

Paraphrasing the definition given in [5], symbolic analysis is "*a formal technique to calculate the behavior or attribute of a system with all its independent variables, dependent variables, and (some of its) parameters represented by symbols.*" In the specific case of the tapered beam, the abscissa x is an independent variable while the mode shape ϕ_r and the eigenvalue β_r are dependent variables. The taper coefficient α is the parameter of interest, and according to the definition, all these quantities must be defined by symbols. If a solution of the untapered ($\alpha = 0$) beam equation (Eq. (5.2)) is known, then the spatial solution of the tapered beam equation (Eq. (5.3)) can be expressed symbolically as the formal power series approximations in α:

$$\phi_r(x) \approx \sum_{n=0}^{N_p} \psi_{r,n}(x)\alpha^n, \text{ and} \tag{5.4}$$

$$\beta_r \approx \sum_{n=0}^{N_p} \mu_{r,n}\alpha^n, \tag{5.5}$$

where N_p is the order of perturbation. Note that the zeroth-order terms in the above series correspond to the $\alpha = 0$ case whose solution is known. In symbolic modeling, all the above quantities, including α, are kept as symbols along with the number of perturbation terms N_p of the approximation series. A major contribution of this chapter is the development and evaluation of an iterative algorithm for the symbolic computation of functions $\psi_{r,n}(x)$ in Eq. (5.4) and coefficients $\mu_{r,n}$ in Eq. (5.5). The symbolic iteration is obtained by substituting Eqs. (5.4) and (5.5) into Eq. (5.3) and collecting terms with equal powers of α to get the following two coupled recursions in $\mu_{r,n}$ and $\psi_{r,n}(x)$, respectively:

$$\mu_{r,n} = -m_0 \sum_{j=1}^{n} \left\langle \frac{2}{l}\left(\frac{x}{l}\right)^{j-1} \frac{d^3\psi_{r,n-j}}{dx^3} \mid \psi_{r,0} \right\rangle_x + m_0 \sum_{j=1}^{n-1} \mu_{r,j} \left\langle \psi_{r,n-j} \mid \psi_{r,0} \right\rangle_x, \text{ and} \tag{5.6}$$

$$\frac{d^4\psi_{r,n}(x)}{dx^4} - \mu_{r,0}\psi_{r,n}(x) = \sum_{j=1}^{n} \left(\frac{2}{l}\left(\frac{x}{l}\right)^{j-1} \frac{d^3\psi_{r,n-j}(x)}{dx^3} + \mu_{r,j}\psi_{r,n-j}(x) \right),$$

where $1 \le n \le N_p$, and the notation $\left\langle f \mid g \right\rangle_x$ denotes the inner product defined by the integral with respect to the spatial variable x:

$$\left\langle f \mid g \right\rangle_x = \int_0^l f(x)g(x)dx.$$

Defining

$$F_{r,j}(x) \triangleq \frac{2}{l}\left(\frac{x}{l}\right)^{j-1} \frac{d^3\psi_{r,n-j}(x)}{dx^3} + \mu_{r,j}\psi_{r,n-j}(x),$$

Eq. (4.15) can be simplified to

$$\frac{d^4\psi_{r,n}(x)}{dx^4} - \mu_{r,0}\psi_{r,n}(x) = S_{r,n}(x), \tag{5.7}$$

where

$$S_{r,n}(x) = \sum_{j=1}^{n} F_{r,j}(x). \tag{5.8}$$

We call the above coupled recursions in Eqs. (5.6) and (5.7) the iterative tapered perturbation solution (ITPS). Its structure calls the following remarks:

1. For each mode r, the coupled recursion of Eqs. (5.6) and (5.7) is initialized with $\mu_{r,0}$ and $\psi_{r,0}(x)$, which are, respectively, the eigenvalue and the eigenfunction of the rth mode of the untapered beam of Eq. (5.2) with its boundary conditions.

2. The differential operator

$$\mathcal{L} \triangleq \frac{d^4}{dx^4} - \mu_{r,0} \tag{5.9}$$

on the left side of Eq. (5.7) is always the same, while only the right side changes from one iteration to another. This should facilitate the solution of Eq. (5.7) by finding Green's function of Eq. (5.9).
3. The right side of Eq. (5.7) requires the use of additional symbols at each iteration. While the third-order derivative terms in Eq. (5.8) can be generated recursively, the symbolic convolution of the sequences $\mu_{r,j}$ and $\psi_{r,j}(x)$ cannot.
4. It is instructive to write the very first step of the coupled recursion. Indeed, for $n = 1$, we have

$$\mu_{r,1} = -m_0 \left\langle \frac{2}{l} \frac{d^3\psi_{r,0}}{dx^3} | \psi_{r,0} \right\rangle_x, \text{ and}$$

$$\frac{d^4\psi_{r,1}}{dx^4} - \mu_{r,0}\psi_{r,1} = \frac{2}{l} \frac{d^3\psi_{r,0}}{dx^3} + \mu_{r,1}\psi_{r,0}.$$

This clarifies the fact that at iteration n, the coefficient $\mu_{r,n}$ has to be computed before the function $\psi_{r,n}(x)$.
5. Finally, while the recursive expressions of Eqs. (5.6) and (4.15) may seem daunting, their derivation is greatly facilitated by the use of the symbolic language of Mathematica. Indeed, the substitution of Eqs. (5.4) and (5.5) into Eq. (5.3) in Mathematica is a straightforward operation, while the collection of equal powers of α to derive the recursive expressions can be readily accomplished using Mathematica's COLLECT operator.

Figure 5.2 shows the various symbolic modules used to construct the series of modal perturbations of the tapered beams. Three main modules parallel the derivation of Eqs. (5.6) and (5.7). They are device initialization, system formulation, and perturbation analysis, all executed symbolically.

1. **Device initialization**: The first step in building the device model is to define the device (*dvc*) in terms of its technology parameters and geometry. Parameters are maintained symbolically in the model with the option of substituting corresponding design values to realize the numeric equivalent of the model. The parameters subject to tapering are then defined in terms of the tapering coefficient α.
2. **System formulation**: Once the tapered device is initialized, the governing system (*sys*) is formulated symbolically using the *system formulation API*, taking into consideration the governing physics of the device and the corresponding equilibrium conditions. Modal analysis is then performed on the system symbolically by employing a separation of the spatial and temporal variables leading to the modal temporal solution (MTS) and modal spatial solution (MSS).

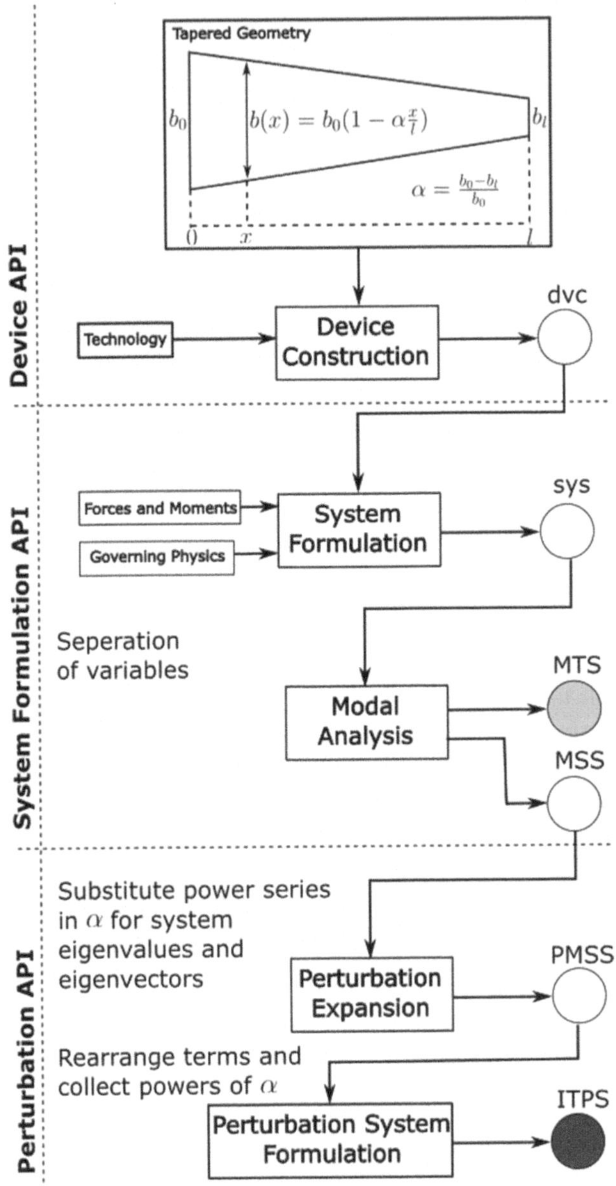

Fig. 5.2 Symbolic modules for constructing device geometry, formulating the device model, and applying perturbation expansion

3. **Perturbation analysis**: Perturbation expansion and analysis are performed symbolically on MSS using the *perturbation API*. This is done by substituting a power series in α for the eigenvalues and eigenfunctions of MSS to produce

```
GetPerturbedSystem[sys_, x_, alpha_, phi_, beta_] :=
Module[{sysOut, sysP, phiT, betaT, C, temp, cIdx},
    sys[xLoc_] = Substitute[sys[xLoc], ϕ[x], Sum[ψ[n][xLoc] α₁^n, {n, N}]];
    sys[xLoc_] = Substitute[sys[xLoc], β, Sum[μ[n] α₁^n, {n, N}]];
    sys[xLoc_] = MultiplyOutSequences[GetLHS[sys[xLoc]]] == 0;
    sysOut[xLoc_] = FindSequenceCoefficient[GetLHS[sys[xLoc]], n] == 0;
    Return[sysOut];
];
```

Fig. 5.3 Mathematica code snippet for symbolic perturbation expansion

the *perturbed MSS* (PMSS). After rearranging the terms and comparing the coefficients of equal powers in α, the iterative tapered perturbed solution (ITPS) is derived.

A Mathematica code snippet is shown in Fig. 5.3 for function **GetPerturbedSystem**, which takes as arguments the modal system **SysX**; the spatial variable **x**; the perturbation parameter, namely, the tapering coefficient α; and the symbols to expand as power series of the perturbation parameter, namely, the mode coefficient β and the mode shape ϕ. It returns a sequence of equations indexed by the integer n. The symbolic nature of this function is best illustrated by the fact that the order of perturbation N_p is not an argument of **GetPerturbedSystem**. This is unlike a numerical function where a numerical value must be assigned to N_p so that the perturbation expansion can be numerically evaluated. See Sect. 5.4. A given term in the symbolic expansion has a recursive dependence on the preceding terms. As we shall see in the next section, this recursion leads to a complete symbolic solution of all terms in the perturbation series for an arbitrary order N_p.

5.4 Algorithm for Symbolic Modal Analysis

This section is devoted to a detailed description of the symbolic iterative algorithm used to derive a full symbolic modal approximation of the tapered beam. The algorithm is represented by the diagram in Fig. 5.4 and comprises the following steps: initialization, eigenvalue computation, Green's function derivation, mode-shape computation, and normalization. We first describe the mathematical content of each of these steps and then end the section with a brief description of the Mathematica implementation of various blocks in Fig. 5.4.

5.4.1 Iteration Initialization

In iteration initialization, we solve for the zeroth-order terms, $\mu_{r,0}$ and $\psi_{r,0}$, in the series expansion of the eigenvalues and eigenfunctions, respectively. To do so,

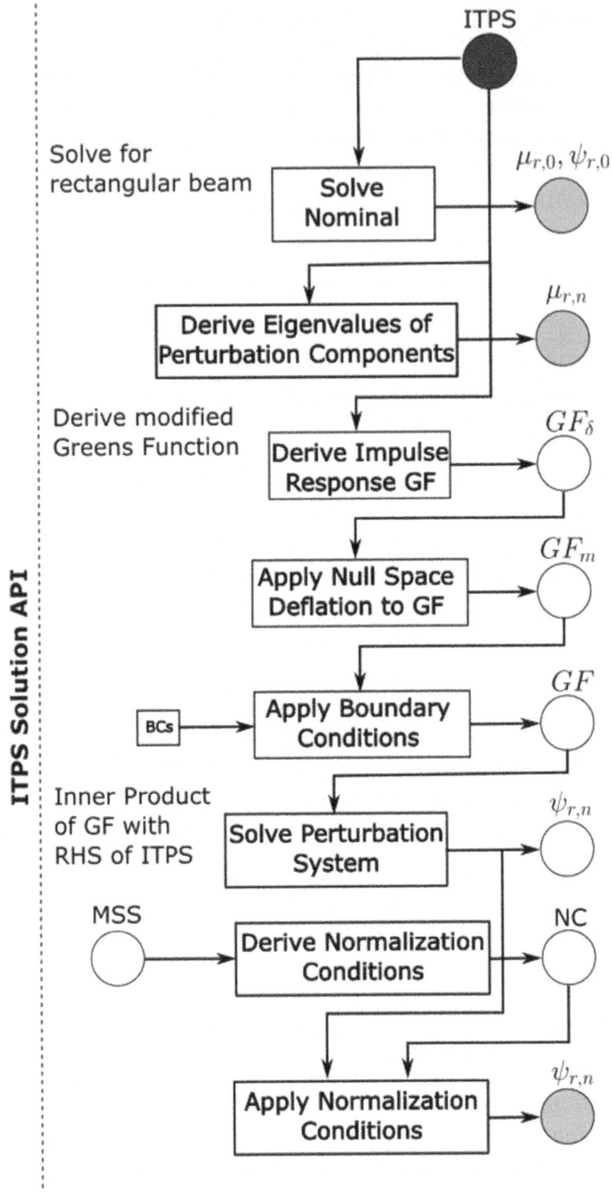

Fig. 5.4 Symbolic modules for solving the automatic perturbation expansion of the tapered beam, including solving for Green's function (GF). The light gray circles indicate the symbolic outputs of the algorithm at each iteration. The left white circles indicate intermediate symbolic results. Note that $\psi_{r,n}$ appears twice, before normalization and after normalization

we solve the beam equation of the nominal case, which is that of the rectangular cantilever beam of Sect. 5.2. The modal solution of this equation is well known [12] with the eigenvalues given by

$$\mu_{r,0} = \left(\frac{\lambda_r}{l}\right)^4,$$

where λ_r is computed as the solution of the characteristic equation $1 + \cos\lambda_r \cosh\lambda_r = 0$. The eigenfunctions are given by

$$\psi_{r,0}(u) = \frac{1}{\sqrt{lm}}(\cosh u - \cos u - \sigma_r(\sinh u - \sin u)), \tag{5.10}$$

where

$$u = \frac{\lambda_r x}{l}, \quad \sigma_r = \frac{\sinh\lambda_r - \sin\lambda_r}{\cosh\lambda_r - \cos\lambda_r}.$$

5.4.2 Symbolic Eigenvalue Computation

In this step, the coefficients $\mu_{r,n}$, $n \geq 1$, of the eigenvalue power series are computed symbolically according to Eq. (5.6). Note that the expression of $\mu_{r,n}$ is obtained by writing the orthogonality condition between $\psi_{r,0}(x)$ and $S_{n,r}(x)$, the loading function on the right side of the $\psi_{r,0}(x)$ differential equation (5.7). This orthogonality condition is a consequence of the fact that the operator $\mathcal{L}$ of Eq. (5.9) is Hamiltonian.

5.4.3 Symbolic Green's Function Derivation

There are two ways to calculate the functions $\psi_{r,n}(x)$ of the rth mode shape from Eq. (4.15). The first is to directly use the variation of parameters on a solution of the homogeneous equation (Eq. (5.10)). This is rather cumbersome and requires repeated symbolic evaluations to get the special solution for each loading function $S_{r,n}(x)$. The second way is to symbolically derive Green's function of Eq. (5.9), also using variation of parameters, and to write the solution for an arbitrary load $S_{r,n}(x)$ as the symbolic inner product of $S_{r,n}(x)$ and Green's function. We have adopted the Green's function approach as it enables the symbolic solution to be pushed all the way to the point where specific design candidates can be evaluated numerically to any given perturbation order.

Green's function is the impulse response of the $\mathcal{L}$ operator in Eq. (5.9), satisfying

$$\mathcal{L}G_r(x,\xi) = \frac{d^4 G_r(x,\xi)}{d\xi^4} - \mu_{r,0} G_r(x,\xi) = \delta(x-\xi). \tag{5.11}$$

Since a homogeneous solution for (5.11) exists, the null space needs to be deflated from Green's function, which amounts to evaluating modified Green's function $G_{r,m}(x,\xi)$, which is a solution to the deflated equation:

$$\mathcal{L}G_{m,r}(x,\xi) = \delta(x-\xi) + \psi_{r,0}(x)\psi_{r,0}(\xi).$$

The symbolic code includes various methods for evaluating $G_{m,r}(x,\xi)$, which is first split into two parts, the impulse response part, $G_{\delta,r}(x,\xi)$, and the deflation part, $G_{0,r}(x,\xi)$ such that $G_{m,r}(x,\xi) = G_{\delta,r}(x,\xi) + G_{0,r}(x,\xi)$. A symbolic implementation of the variation-of-parameters solver is used to solve the deflated part $\mathcal{L}G_{0,r}(x,\xi) = \kappa\psi_{r,0}(x)\psi_{r,0}(\xi)$ as

$$G_{0,r}(x,\xi) = \frac{l^2}{8\lambda_r^4}\sqrt{lm}\ \psi_{r,0}(x)\Big(3l\sqrt{lm}\ \psi_{r,0}(\xi) + 2\lambda_r\xi\sigma_r\sqrt{lm}\ \nu_{0,r}(\xi)\Big),$$

where

$$\nu_{0,r}(\xi) = \frac{1}{\sqrt{lm}}\left(\cosh\frac{\lambda_r\xi}{l} - \cos\frac{\lambda_r\xi}{l} - \frac{1}{\sigma_r}\left(\sinh\frac{\lambda_r\xi}{l} + \sin\frac{\lambda_r\xi}{l}\right)\right).$$

As for the impulse response part $\mathcal{L}G_{\delta,r}(x,\xi) = \delta(x-\xi)$, sinusoidal and hyperbolic trial functions are used as in the case of the nominal rectangular beam [12]. The solution $G_{\delta,r}(x,\xi)$ consists of two segments joined through a discontinuity at the impulse point ($x = \xi$) using boundary and continuity conditions as shown in Fig. 5.5. These conditions are defined at both ends of the beam and at $x = \xi$.

The final form of $G_{\delta,r}(x,\xi)$ is rather cumbersome, with many more terms than can be handled cleanly. To describe Green's function compactly, various sets of multipliers of the trial bases

$$\boldsymbol{\Psi}_r(x) = [\cos(\lambda_r x/l), \cosh(\lambda_r x/l), \sin(\lambda_r x/l), \sinh(\lambda_r x/l)],$$

and various flavors of mode shapes have to be symbolically coded. As for Green's function $G_{\delta,r}(x,\xi)$ of the impulse response, it has a component $G^{(a)}_{\delta,r}(x,\xi)$ that is scaled by the free coefficient c_1 and a component $G^{(b)}_{\delta,r}(x,\xi)$ that is not. The final modified Green's function of the rth mode shape is then expressed as

$$G_{m,r}(x,\xi) = c_1 G^{(a)}_{\delta,r}(x,\xi) + G^{(b)}_{\delta,r}(x,\xi) + G_{0,r}(x,\xi). \tag{5.12}$$

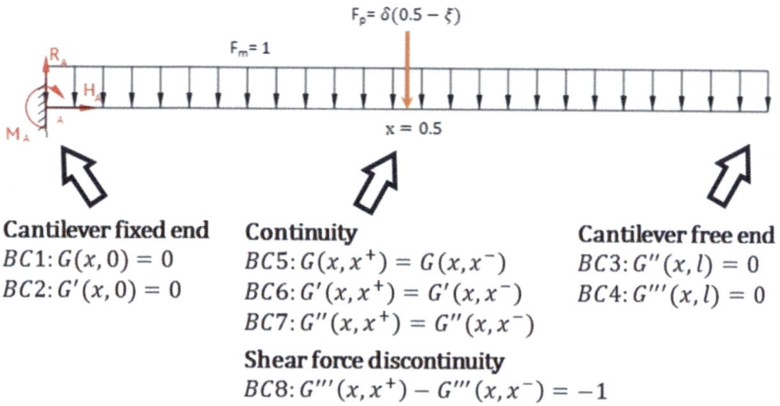

Fig. 5.5 Boundary conditions of Green's function $G(x, \xi)$ at the beam ends and the conditions imposed by the impulse. The derivatives shown are with respect to ξ [3]

The reader is referred to Chap. 4 for the full symbolic expression of Green's function.

5.4.4 *Symbolic Mode-Shape Computation*

The general solution for the nth perturbed component of the rth mode shape is obtained by convolving Green's function $G_{m,r}(x, \xi)$ with the forcing term of Eq. (4.15):

$$\psi_{r,n}(x) = \int_0^l G_{m,r}(x, \xi) S_{r,n}(\xi) d\xi.$$

When using the Green's function components described in (5.12), the above expression can be written as

$$\begin{aligned}\psi_{r,n}(x) =& c_1 \int_0^l G^{(a)}_{\delta,r}(x, \xi) S_{r,n}(\xi) d\xi \\ &+ \int_0^l (G^{(b)}_{\delta,r}(x, \xi) + G_{0,r}(x, \xi)) S_{r,n}(\xi) d\xi.\end{aligned}$$

This is the complete symbolic description except for the determination of the free coefficient c_1. This we will do in the normalization step of the following subsection.

5.4.5 Symbolic Normalization

The normalization conditions for each function $\psi_{r,n}(x)$ is symbolically evaluated using the orthogonality of the mode shapes $\phi_r(x)$ with respect to the mass density $m(x)$:

$$\int_0^l \phi_r(x)m(x)\phi_s(x)dx = \delta_{rs}.$$

The mode shapes $\phi_r(x)$ and $\phi_s(x)$ are expanded using Eq. (5.4), taking into account the linear distribution of $m(x)$. Collecting equal powers in α from the resulting expansion, one can write for the $\psi_{r,n}(x)$ functions:

$$\begin{aligned} \int_0^l \psi_{r,0}(x)\psi_{s,n}(x)dx =& \frac{1}{2l}\int_0^l x\sum_{i=0}^{n-1}\psi_{r,i}(x)\psi_{s,n-1-i}(x)dx \\ &-\frac{1}{2}\int_0^l \sum_{i=1}^{n-1}\psi_{r,i}(x)\psi_{s,n-i}(x)dx. \end{aligned} \tag{5.13}$$

The normalization condition can then be written by taking $s = r$ in the previous equation and expressing the integrals as inner products:

$$\left\langle \psi_{r,0} \mid \psi_{r,n} \right\rangle_x = P_{r,n}, \tag{5.14}$$

where

$$\begin{aligned} P_{r,n} =& \frac{1}{2l}\sum_{i=0}^{n-1}\left\langle x\psi_{r,i} \mid \psi_{r,n-1-i}\right\rangle_x \\ &-\frac{1}{2}\sum_{i=1}^{n-1}\left\langle \psi_{r,i} \mid \psi_{r,n-i}\right\rangle_x . \end{aligned}$$

Since

$$\begin{aligned} \psi_{r,n}(x) =& c_1\langle G_{\delta,r}^{(a)}(x,\xi) \mid S_{r,n}(\xi)\rangle_\xi \\ &+ \langle G_{\delta,r}^{(b)}(x,\xi) + G_{0,r}(x,\xi) \mid S_{r,n}(\xi)\rangle_\xi, \end{aligned}$$

its substitution into (5.14) yields a symbolic formulation for the free coefficient:

$$c_1 = \frac{P_{r,n} - \langle \psi_{r,0}(x) \mid \langle G_{\delta,r}^{(b)}(x,\xi) + G_{0,r}(x,\xi) \mid S_{r,n}(\xi)\rangle_\xi\rangle_x}{\langle \psi_{r,0}(x) \mid \langle G_{\delta,r}^{(a)}(x,\xi) \mid S_{r,n}(\xi)\rangle_\xi\rangle_x}. \tag{5.15}$$

Note that all the relationships from Eq. (5.13) through Eq. (5.15) are derived symbolically using Mathematica, and there is no need for any numerical evaluation. This fully symbolic, iterative derivation of the perturbation solution using symbolic Green's functions is implemented in Mathematica in line with the ITPS diagram of Fig. 5.4. We will briefly describe this implementation in the next subsection.

5.4.6 Symbolic Implementation

The symbolic implementation of the steps described in the previous subsections closely follows the diagram of Fig. 5.4. Note that the nominal solution to initialize the zeroth-order terms $\mu_{r,0}$ and $\psi_{r,0}(x)$ is executed only once, while the subsequent steps to compute $\mu_{r,n}$ and $\psi_{r,n}(x)$ are repeated for $n \in \{1, \ldots, N_p\}$ where N_p is the perturbation order. This order is implicit in the symbolic iterative solution but becomes explicit when the symbolic perturbation model is handed to the numerical evaluator.

A snippet of the Mathematica code of the symbolic solution is given in Fig. 5.6. The snippet contains the derivation of the modified Green's function using the variation-of-parameters method. Firstly, the trial functions of the nominal case are defined and then the fourth-order differential operator is specified. Subsequently, the loading functions are defined and used to compute the two parts of the modified Green's function: the impulse response and the particular solution used in the deflation. As mentioned in Sect. 5.4.4, the contribution to the mode shape is the result of the symbolic inner product of the sum of these two responses and the loading function on the right side of Eq. (4.15), which is calculated from previous iterations. This inner product and the normalization step are implemented in other Mathematica modules.

```
In[ ]:= trials[x_] = {Cos[xλ/l1], Cosh[xλ/l1], Sin[xλ/l1], Sinh[xλ/l1]};

LL[x_] = GetLHS[ITPS[x]]
(*LL[y_,x_]=D[y[x],{x,4}]-(λ/l1)^4 y[x];*)
F1[ξ_] = DiracDelta[ξ - x];
F0[ξ_] = -m0 ϕ0x[x] ϕ0[ξ];

G0 = SolveUsingVariationOfParameters[F0, trials];

Gδ = GetGreenFunction[LL[x]];

G = Gδ[x] + G0[x];
```

Fig. 5.6 Mathematica code snippet implementing the iterative solution of the perturbation expansion

The actual evaluation of the device's modal and dynamic responses is deferred to the numerical solvers.

5.5 Model Evaluation Modules

The numerical evaluation of the symbolic model closely follows the steps described in the previous section for symbolic model generation. The various numerical modules for model evaluation are shown in Fig. 5.7. The modules consist of two groups. The first group is responsible for the numerical evaluation of mode-shape approximations using the perturbation expansion. These modules are accessed through the *numerical tapered beam API.* The second group is responsible for the computation of the model's temporal response under various external inputs. These modules are accessed through the *numerical ODE solver*. Both groups are implemented with MATLAB scripts.

5.5.1 Numerical Evaluation of Perturbation Expansions

About the top of Fig. 5.7, the numerical evaluations of the perturbation expansions of eigenvalues and mode shapes trace the steps to build the symbolic expansion. Numerical values describing the geometry of the beam and materials are substituted for their symbols, and the eigenvalues and mode shapes of the rectangular beam are computed. Then the numerical iteration is entered to compute the terms of the perturbation expansions of β_r and $\phi_r(x)$. The iteration is terminated at a given expansion order N_p. Resonance frequencies are evaluated from the eigenvalues as $\omega_r = \sqrt{\beta_r Y I_0/m_0}$ and used for the spectral analysis of the beam. The MATLAB script of this numerical solution is shown in Fig. 5.8

5.5.2 Computation of Temporal Response

The evaluated symbolic model of the modal spatial solution (MSS) along with the modal temporal solution (MTS) is combined to derive a lumped-parameter model of the beam [11]. In the case of TPEH, the procedure results in a full electromechanical model that can be simulated using an ordinary differential equation solver such as ODE in MATLAB. In the case of CMOS-MEMS design, a SPICE solver can also be used. Note that the order of the lumped model depends on the order of the perturbation expansion N_p. The computation of the temporal response is implemented as MATLAB script in the *numerical tapered beam API* which, in tandem with MATLAB's ODE solver, provides an environment for fast-turnaround design space exploration of the tapered beam.

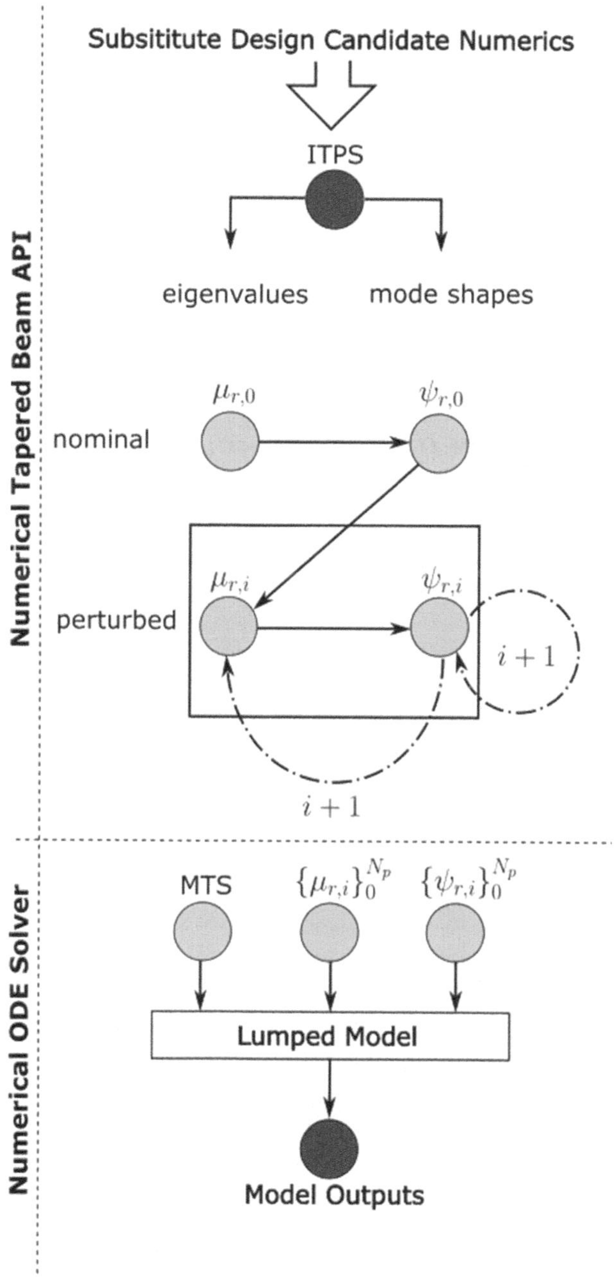

Fig. 5.7 Numerical modules for evaluating the symbolic perturbation expansion. An ODE solver is used to compute the temporal response of the model

```
function [Phi,f] =  modalAnalysis(dvcInput,r,N) % h: API
% Computes resonance frequencies and mode shapes for mode r. Uses
% N perturbation components to solve for the tapered beam case.
% Setup Device Geometry
x = dvcInput.x; alpha = dvcInput.alpha; l = dvcInput.l;
m0 = dvcInput.m0; YI0 = dvcInput.YI0;
% Pre allocate
phi = zeros(size(x,1),N+1);
G = zeros(size(x,1),size(x,1),N+1);
D3Phi = zeros(size(x,1),N+1);
mu = zeros(1,N);
% Nominal mode shape and eigenvalue
[phi(:,1),~,lambda] = modalAnalysis_CFR(dvcInput,r);
mu(1) = (lambda/l).^4; % System eigenvalue for rth mode
% Get Green's function
[G0, Ga, Gb, xi] = getGF_CFT(dvcInput,lambda);
% Evaluate perturbed components
J=0:N; Xp = 2/l*repmat(x/l,[1,N+1]).^repmat(J-1,size(x));
% Evaluate terms needed in evaluating perturbed components in the loop
D3Phi(:,1) = h.u.derivativeN(x,phi(:,1),3,'fornberg');
% Loop to evaluate perturbed eigenvalues and mode shape components
j = 0:N;
for n=1:N
    dTermSum = sum(Xp(:,h.u.getIdx(j,1,n))...
        .*D3Phi(:,h.u.getIdx(n-h.u.getSubI(j,1,n))),2);
    % Compute eigenvalue of the nth component
    mu(h.u.getIdx(n)) = -m0*(h.u.getInnerProduct(dTermSum,phi(:,1),x)...
        + h.u.ifTF(n>1,mu(h.u.getIdx(j,1,n-1))*h.u.getInnerProduct(...
        phi(:,1),phi(:,h.u.getIdx(n-h.u.getSubI(j,1,n-1))),x),0));
    % Compute RHS F
    F = dTermSum + phi(:,h.u.getIdx(n-h.u.getSubI(j,1,n)))...
        *mu(h.u.getIdx(j,1,n))';
    % Compute next perturbed component
    phiGa = h.u.getInnerProduct(F,Ga',xi);
    phiGb = h.u.getInnerProduct(F,Gb',xi);
    phiG0 = h.u.getInnerProduct(F,G0',xi);
    % Normalize GF
    c1 = getNormalizationConstantForGF_CFT...
        (dvcInput,phi,phiGa,phiGb,phiG0,j,n);
    phi(:,h.u.getIdx(n))= c1*phiGa+phiGb+phiG0;
    G(:,:,h.u.getIdx(n))=c1*Ga+Gb+G0'; % for records only
    % Compute next derivative
    D3Phi(:,h.u.getIdx(n)) = h.u.derivativeN(x,...
        phi(:,h.u.getIdx(n)),3,'fornberg');
end
Phi = (alpha.^(0:N)*phi')';
Beta = alpha.^(0:N)*mu';
f = (1/(2*pi))*sqrt(YI0*Beta/m0);
end
```

Fig. 5.8 Numerical solution using a MATLAB script

5.6 Modal Analysis Validation Against MEMS+ Results

A tapered beam model was constructed using MEMS+ with the material properties shown in Table 5.1. Note that two molybdenum layers were also defined in MEMS+ to model the electrodes. Small beams were carefully put together in series to form a larger beam with the desired mechanical node density for a detailed analysis of the beam model and its comparison with the analytical model. The proposed model works equally well for a MEMS scale structure and a larger structure for as long as the small deflection assumption is not grossly violated. Here, a larger structure is considered for our model validation owing to the availability of benchmark data published in [13, 14].

The structure of the beam was parameterized with the taper factor $\alpha = \frac{b_0 - b_l}{b_0}$. A rectangular beam is modeled by setting $\alpha = 0$, whereas $\alpha = 1$ corresponds to a triangular beam. A true triangular beam, however, cannot be modeled using MEMS+ because of its inability to specify zero widths so instead a taper factor close to 1 such as 0.99 was chosen to model a triangular beam.

5.6.1 *Resonance Frequency Estimation Results*

Resonance frequencies of the first three modes were computed using MEMS+ for a range of α values and compared against the proposed analytical model. The results are shown in Fig. 5.9. The percentage errors shown in the second column are computed with MEMS+ solution as the reference. We notice an increasing trend in the percentage error as the taper in the beam increases.

A percentage error *versus* N_p view in Fig. 5.10 shows an exponentially decreasing trend as the number of perturbation terms increases. The errors are higher for

Table 5.1 Geometric and material parameters of test energy harvester [12]

Parameter name	Symbol	Value	Units
Beam length	l	0.1	m
Beam width at base	b_0	0.02	m
Substrate thickness	h_s	5×10^{-4}	m
PZT thickness	h_p	4×10^{-4}	m
Mo electrode thickness	h_e	2×10^{-7}	m
Substrate Young's modulus	Y_s	100	GPa
PZT Young's modulus	Y_p	66	GPa
Mo Young's modulus	Y_e	329	GPa
Substrate mass density	ρ_s	7165	kg/m^3
PZT mass density	ρ_p	7800	kg/m^3
Mo mass density	ρ_e	10220	kg/m^3
PZT piezoelectric constant	d_{31}	−190	pm/V
PZT permittivity	ε_{33}^S	15.93	nF/m

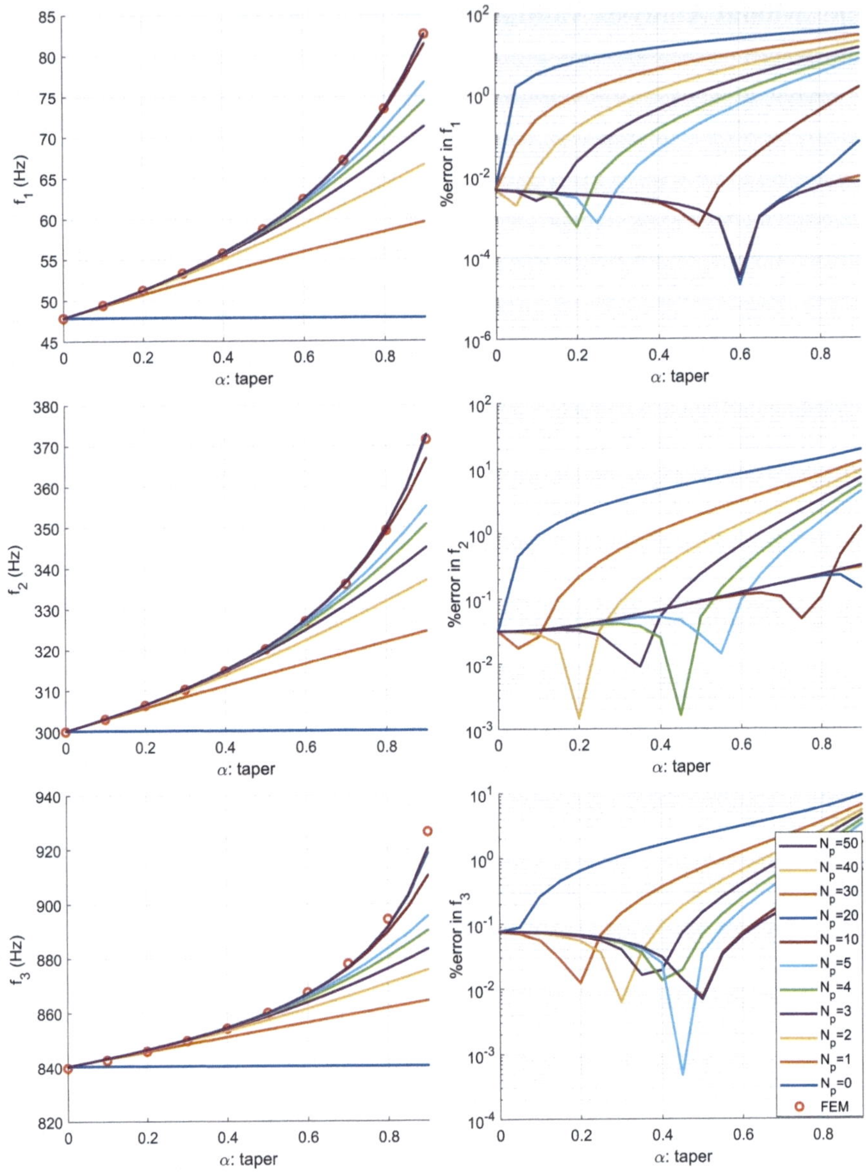

Fig. 5.9 Resonance frequency comparison with MEMS+. f_r and % error in f_r of the first three modes (top to down) presented as α are varied for a list of N_p values

higher tapers and the error values in general saturate to a floor of around 0.001% for f_1 estimates. Both Figs. 5.9 and 5.10 indicate that the analytical model favors a taper ratio of 0.6 for the computation of the first mode resonance frequency for 20 perturbation components or more. Likewise, different taper factors favor different

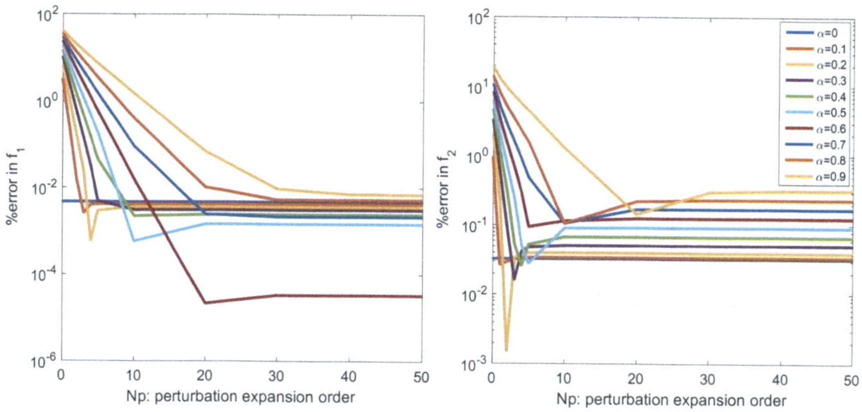

Fig. 5.10 Resonance frequency error as a function of N_p

N_p for their resonance frequency estimates evidenced by the dips in the error curves. We can exploit this phenomenon to assign the optimal N_p for different α values when computing resonance frequencies for design variants with different tapers in a design optimization loop.

5.6.2 *Mode-Shape Estimation Results*

The mode shapes estimated from the perturbation expansion solution presented here were also compared against the FEM solution obtained from MEMS+. Figure 5.11 shows results for the first three mode shapes of beams with $\alpha = 0.1$, $\alpha = 0.5$, and $\alpha = 0.9$. The number of samples needed when estimating a mode shape depends on the mode number. For higher modes, a denser spatial sampling is required for reasonable results. These results are obtained for 101 samples along the length of the beam. Our analysis indicates reasonable results for the first three mode shapes for as few as 11 samples. We notice that whereas the first mode shape does not diverge much from the rectangular solution, the higher modes are quite different as the taper factor is increased. It therefore requires a higher number of perturbation components for reasonable estimates of the higher mode shapes for beams with higher taper angles.

Figure 5.12 shows the percentage errors between the first three mode shapes obtained from the proposed semi-analytical algorithm and those obtained from MEMS+ for taper factors of $\alpha = 0.1$, 0.5, and 0.9 plotted against normalized distance from the anchor along the length of the beam for different numbers of perturbation expansion components used for the estimation. Though the errors are generally low, the occasional peaks can be explained by a slight offset between

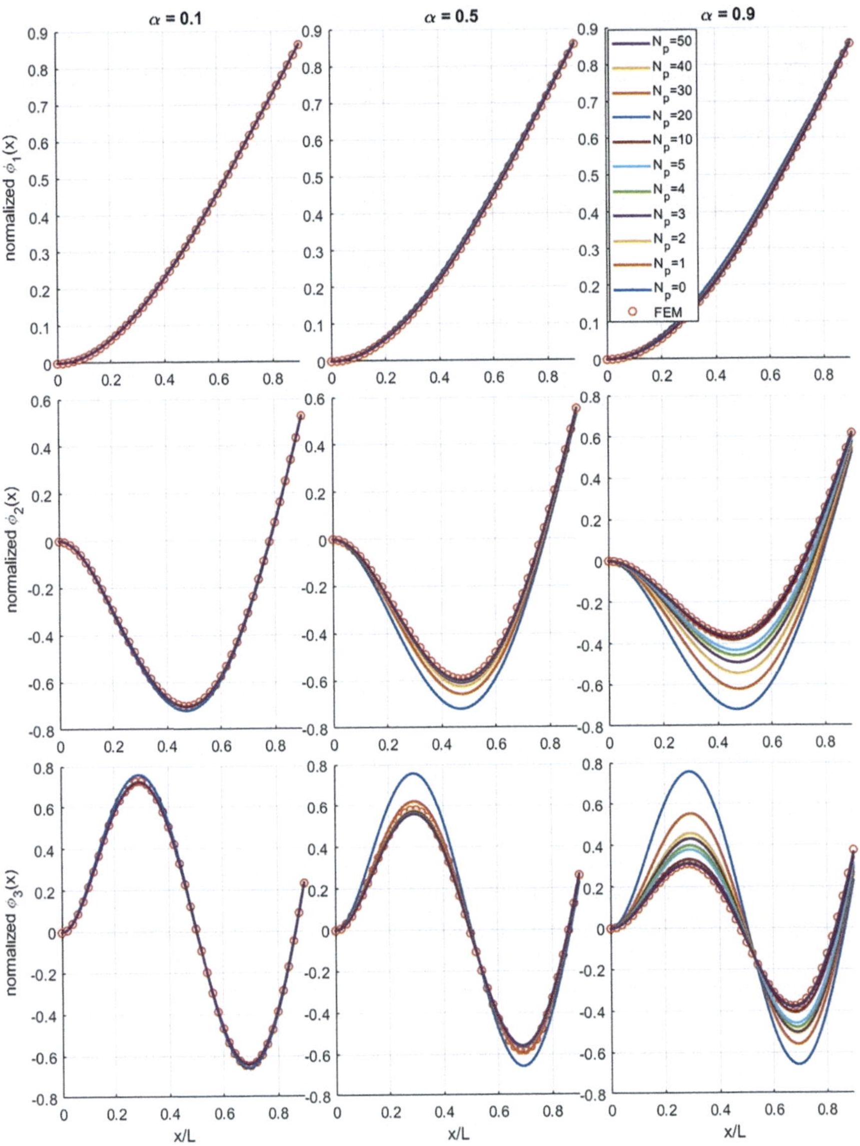

Fig. 5.11 Mode-shape estimates of the first three modes and their comparison with MEMS+ results for $\alpha = 0.1$, $\alpha = 0.5$, and $\alpha = 0.9$ for a list of N_p

the spatial sampling implemented by the two methodologies and the changes in curvature in the mode shapes for the higher mode.

Finally, we present the sum squared of the mode-shape errors plotted against α in Fig. 5.13 showing that the sse generally increases as the taper ratio α increases and decreases with N_p.

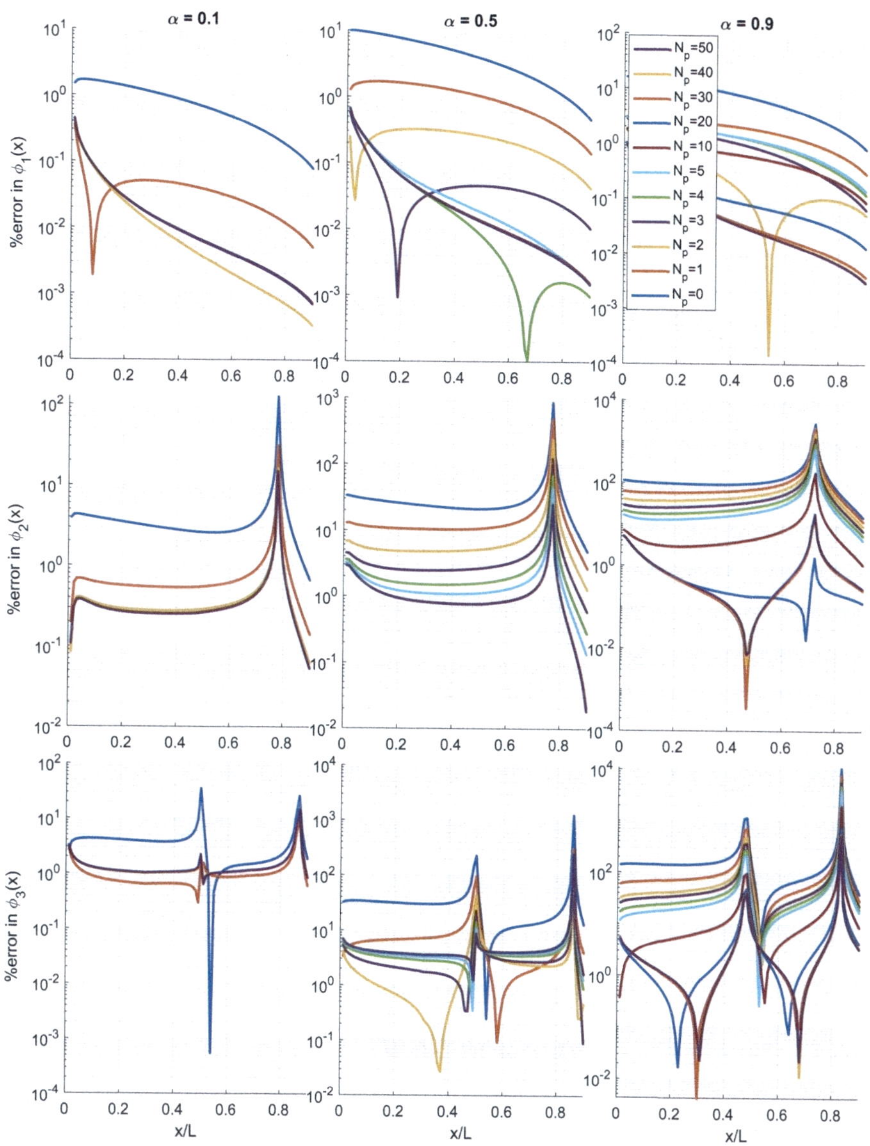

Fig. 5.12 Percentage error in $\phi_1(x)$ compared to MEMS+ FEM solution using 101 samples along the beam for $\alpha = 0.1,\ 0.5,$ and 0.9 for different N_p plotted against the normalized distance along the beam length

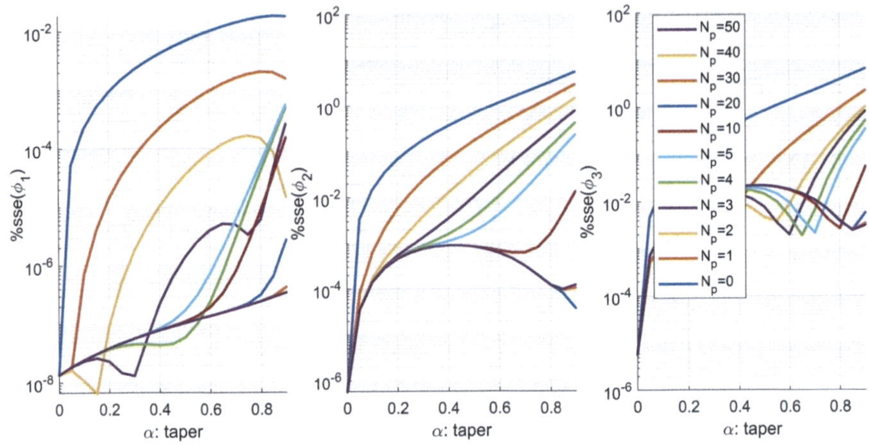

Fig. 5.13 Sum squared errors of the first three modes plotted against α for different N_p

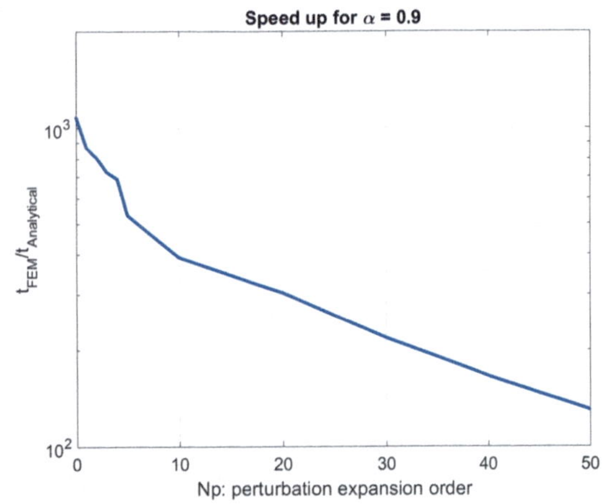

Fig. 5.14 Modal analysis timing speedup compared to FEM results *versus* N_p for a structure with $\alpha = 0.9$

5.6.3 Timing Comparison with FEM

We have used MEMS+ as the benchmark FEM tool for model validation. MEMS+ is a network-based, fast FEM software that uses a minimal number of spatial nodes for its analyses. When using a comparable spatial sampling to enable sample-to-sample comparison of the mode shapes, we notice a remarkable increase in computational efficiency when using our proposed semi-analytical tapered beam model. The speedup when using 20 perturbation components is as high as $\times 300$ and up to $\times 700$ when using 4 perturbation components. The simulation speedup when using fewer number of spatial samples is also significant and is explored in more detail in the next chapter (Fig. 5.14).

5.7 Conclusions

In this chapter, we have presented a novel method for deriving symbolic models of tapered beams used in MEMS. The resulting models amount to symbolic approximations of modal solutions to the partial differential equations that describe the beam's spatial and temporal behavior. The symbolic approximation is based on a perturbation expansion in the power series of the taper coefficient. The reference perturbation case is that of an untapered beam. The perturbation expansion terms can be symbolically solved by a rigorous iterative procedure that includes the symbolic evaluations of Green's functions. One of the main contributions of this work is the development of a special library of symbolic Mathematica modules to help construct a term-by-term perturbation expansion and build a series of symbolic eigensolutions for the beam. The symbolic expansion can be evaluated numerically using MATLAB scripts. The numerical results demonstrate the tunable accuracy of the symbolic solution as a function of the taper coefficient and the number of expansion terms. Although only a symbolic model of the mechanical component is presented, the entire electromechanical system of, say, a TPEH can also be modeled [3]. Finally, note that this symbolic method is generic and can be applied to other physical domains, including optics, acoustics, and fluid dynamics.

References

1. Jessy Baker, Shad Roundy, and Paul Wright. "Alternative geometries for increasing power density in vibration energy scavenging for wireless sensor networks". In: *3rd international energy conversion engineering conference*. 2005, p. 5617.
2. *MEMS+ User's Guide*. Coventor Inc. 2016.
3. Wajih U. Syed and Ibrahim M. Elfadel. "Vibration Model of a Prismatic Multilayered Tapered Cantilever Using Perturbation Analysis". In: *Journal of Sound and Vibration* (2019).
4. G.G.E. Gielen, H.C.C. Walscharts, and W.M.C. Sansen. "ISAAC: a symbolic simulator for analog integrated circuits". In: *IEEE Journal of Solid-State Circuits* 24.6 (1989), pp. 1587–1597.
5. G. Gielen, P. Wambacq, and W.M. Sansen. "Symbolic analysis methods and applications for analog circuits: a tutorial overview". In: *Proceedings of the IEEE* 82.2 (1994), pp. 287–304.
6. P. Wambacq et al. "Symbolic analysis of large analog circuits using a sensitivity-driven enumeration of common spanning trees". In: *IEEE Transactions on Circuits and Systems II: Analog and Digital Signal Processing* 45.10 (1998), pp. 1342–1350.
7. Guoyong Shi. "Topological Approach to Symbolic Pole-Zero Extraction In-corporating Design Knowledge". In: *IEEE Transactions on Computer-Aided Design of Integrated Circuits and Systems* 36.11 (2017), pp. 1765–1778.
8. Guoyong Shi. "Generating the Closed-Form Second-Order Characteristics of Analog Differential Cells by Symbolic Perturbation". In: *IEEE Transactions on Circuits and Systems I: Regular Papers* 65.9 (2018), pp. 2939–2950.
9. Hamed Aminzadeh, Alfio Dario Grasso, and Gaetano Palumbo. "A Methodology to Derive a Symbolic Transfer Function for Multistage Amplifiers". In: *IEEE Access* 10 (2022), pp. 14062–14075.

10. Wajih Umar Syed and Ibrahim M. Elfadel. "Symbolic Modeling of Tapered Beams in Piezoelectric Energy Harvesters". In: *2023 Symposium on Design, Test, Integration and Packaging of MEMS and MOEMS (DTIP)*. Valetta, Malta, 2023, pp. 1–4.
11. Wajih Umar Syed, Alabi Bojesomo, and Ibrahim M Elfadel. "Electromechanical model of a tapered piezoelectric energy harvester". In: *IEEE Sensors Journal* 18.14 (2018), pp. 5853–5862.
12. Alper Erturk and Daniel J Inman. "A distributed parameter electromechanical model for cantilevered piezoelectric energy harvesters". In: *Journal of vibration and acoustics* 130.4 (2008), p. 041002.
13. Alper Erturk and Daniel J Inman. "Issues in mathematical modeling of piezoelectric energy harvesters". In: *Smart Materials and Structures* 17.6 (2008), p. 065016.
14. Yaowen Yang and Lihua Tang. "Equivalent circuit modeling of piezoelectric energy harvesters". In: *Journal of Intelligent Material Systems and Structures* (2009), pp. 2223–2235.

Chapter 6
Lumped-Parameter Modeling

Due to high fabrication costs, the microfabrication industry is extremely sensitive to design failures. An iterative design scenario requires working with mathematical and computational models of the device. Numerical simulation tools for MEMS devices that use finite element models have matured over time and have become standard to verify device performance against design specifications [1]. Such tools are now an integral part of the design flow of MEMS devices. Given a list of specifications, computational models for the device are prepared that are simulated to confirm the behavior of the device with respect to the design specifications. The design problem can become difficult to solve when the design space increases with the multiple degrees of freedom, while the design objective also increases in its dimensionality. In such cases, an optimization script is necessary to iterate over the design space and find the best design, while the models need to be as computationally efficient as possible. Accurate and computationally efficient models of the popular MEMS devices are important for developing and improving MEMS CAD modules. Where analytical solutions are possible, they are preferred over numerical FEA solutions due to their computational efficiency and algorithmic and mathematical structure. Structured closed-form analytical solutions enable deep insight into model dynamics that helps to develop more powerful analysis techniques. Efficient solutions to analytical models of the tapered beam structure are hard to derive, as their direct solutions are expressed in terms of special functions such as Meijer G and hypergeometric functions [2].

In this chapter, we present a complete electromechanical model of a tapered piezoelectric energy harvester based on the semi-analytical vibrational model of the tapered cantilever derived in Chap. 4. We derive closed-form steady-state solutions for output voltage, power, and maximum tip deflection of the TPEH for harmonic inputs. We formulate the state-space system of the TPEH to facilitate further analyses, including transient analysis. Finally, we convert the electromechanical dynamic model to an equivalent electrical circuit model (ECM) of the device that can be readily simulated, along with the power conversion and signal conditioning circuitry,

W. U. Syed, I. (Abe) M. Elfadel, *Tapered Beams in MEMS*,
https://doi.org/10.1007/978-3-031-66391-8_6

using commercial CAD environments for circuit design and analysis. The results show that the proposed model can accurately predict the performance of the tapered energy harvester as an ECM extracted from the device's finite element model and can do so $\times 15$ faster for 10 perturbation expansion terms when compared to a reduced-order FEM solution and $\times 150$ faster compared to a regular FEM solution.

6.1 Full Electromechanical Model Formulation

Having derived the mode shapes and the resonance frequencies of the different modes of vibration of the tapered beam in Chap. 4, we can now proceed to derive the modal response equation and the electrical circuit equation with electromechanical coupling.

6.1.1 The Governing System

The damped structural governing equation is derived in Chap. 4 for the TPEH and is shown in Eq. (4.11). Figure 6.1 shows the geometry of the tapered multilayered cantilever beam used to develop the TPEH model. The beam width at the fixed end is b_0, and at the free end, a distance L from the anchor, the width is b_L. The total height of the structure is t, whereas the height of each layer is indexed by the number of layers. The figure illustrates a structure with a silicon (Si) structural base, an aluminum nitride (AlN) piezoelectric layer sandwiched between two molybdenum (Mo) layers that act as charge collecting electrodes, and an oxide (SiO_2) layer to electrically isolate the electrodes and contact pads.

The governing equation of motion for such a tapered Euler–Bernoulli beam when subjected to viscous and strain rate damping can be obtained using Newtonian or Hamiltonian approaches [3]. We use force balancing to set up the governing equation of the tapered PEH which can be written as [4]

$$\frac{\partial^2 M(x,t)}{\partial x^2} + c_a \frac{\partial w_{rel}(x,t)}{\partial t} + m(x) \frac{\partial^2 w_{rel}(x,t)}{\partial t^2} = -m(x) \frac{\partial^2 w_b(x,t)}{\partial t^2} - c_a \frac{\partial w_b(x,t)}{\partial t}, \tag{6.1}$$

where $M(x,t)$ is the internal bending moment, c_a is the viscous air damping coefficient, $m(x)$ is the mass per unit length of the beam at x, and $w_{rel}(x,t)$ is the transverse beam deflection relative to its base, whereas $w_b(x,t) = \gamma_g(x)g(t) + \gamma_h(x)h(t)$ is the base motion with $g(t)$ being the translational motion and $h(t)$ the angular motion and $\gamma_g(x)$ and $\gamma_h(x)$ the corresponding displacement functions. For a cantilever beam $\gamma_g(x) = 1$ and $\gamma_h(x) = x$ so that $w_b(x,t) = g(t) + xh(t)$.

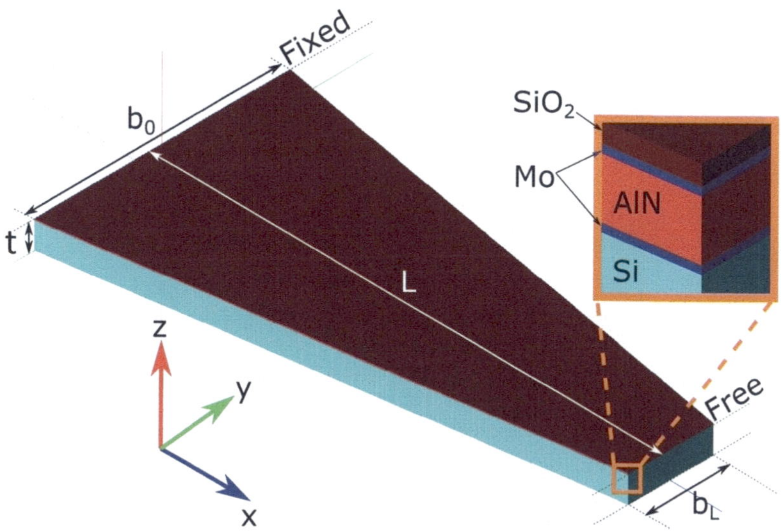

Fig. 6.1 Tapered PEH geometry

6.1.1.1 Internal Moment Evaluation

The total internal moment, $M(x,t)$, is given as

$$M(x,t) = -\sum_{i=1}^{N_L} \int_{z_{i-1}}^{z_i} T_i(x,t) b(x) z dz + M_d(x,t), \tag{6.2}$$

where $M_d(x,t)$ is the strain rate damping bending moment and z_i is the z-location of the ith layer relative to the neutral axis. The stress–strain relationship for the substructure (s) and piezoelectric (p) layers is obtained from the piezoelectric constitutive equations [5]:

$$\begin{aligned} T_s &= Y_s S_s, \text{ and} \\ T_p &= Y_p (S_p - d_{31} E_3), \end{aligned}$$

where T is the stress, S is the strain, Y is Young's modulus, d_{ij} is the piezoelectric constant, and E_i is the electric field. The subscripts s and p represent the substructure and piezoelectric layer indices, respectively, whereas the i and j subscripts represent directions where 1 is along the x-axis, 2 is along the y-axis, and 3 is along the z-axis. Here we assume that in the dominant stress to the electric field, coupling is for the case when the electric polarization occurs in the z-direction (3) across the piezoelectric layer and the stresses develop in the x-direction (1) along the beam length. We use the small deflection bending strain definition

$S(x, z, t) = -z\frac{\partial^2 w_{rel}(x,t)}{\partial x^2}$ for the structural and piezoelectric layers. Moreover, allowing for electrodes of arbitrary length starting from the fixed end, we introduce the Heaviside function ($H(\cdot)$) to define the extent of electrodes over the beam to a point L_e distance from the fixed end. We also consider the contribution of strain rate damping to elaborate the internal moment in Eq. (6.2) as

$$
\begin{aligned}
M(x,t) =& YI(x)\frac{\partial^2 w_{rel}(x,t)}{\partial x^2} \\
&+ \theta(x)v(t)\left(H(x) - H(x - L_e)\right) + c_s I(x)\frac{\partial^3 w_{rel}(x,t)}{\partial x^2 \partial t},
\end{aligned}
$$

where $YI(x) = \sum_{i=1}^{N} YI_i(x)$ with $YI_i(x) = Y_i b(x) \int_{z_i}^{z_{i+1}} z^2 dz$, $v(t) = E_3 h_p$ is the voltage across the piezoelectric layer, and $\theta(x) = -Y_p d_{31} b(x) z_{pc}$ is the coupling factor with z_{pc} the distance of the center of the piezoelectric layer from the neutral axis, c_s is the equivalent strain rate damping coefficient, and $I(x)$ is the homogenized moment of inertia of the beam cross section at x. We identify the three terms on the right-hand side as the structural mechanical bending moment, $M_m(x,t)$; the piezoelectric coupling moment, $M_e(x,t)$; and the strain rate damping bending moment, $M_d(x,t)$. Next, we take the second derivatives of each of the three components of $M(x,t)$ with respect to x before we substitute it in Eq. (6.1).

6.1.1.2 Applying Equilibrium Condition

The force density caused by the transverse mechanical bending moment is evaluated by taking the second derivative of $M_m(x,t)$ with respect to x. We use the product rule to get

$$
\begin{aligned}
\frac{\partial^2 M_m(x,t)}{\partial x^2} =& YI(x)\frac{\partial^4 w_{rel}(x,t)}{\partial x^4} + \frac{\partial^2 YI(x)}{\partial x^2}\frac{\partial^2 w_{rel}(x,t)}{\partial x^2} \\
&+ 2\frac{\partial YI(x)}{\partial x}\frac{\partial^3 w_{rel}(x,t)}{\partial x^3}.
\end{aligned}
\tag{6.3}
$$

The force density from the piezoelectric bending moment is similarly derived as

$$
\begin{aligned}
\frac{\partial^2 M_e(x,t)}{\partial x^2} =& v(t)\left(\frac{d^2\theta(x)}{dx^2}\left(H(x) - H(x - L_e)\right)\right. \\
&\left. +\theta(x)\left(\frac{d\delta(x)}{dx} - \frac{d\delta(x - L_e)}{dx}\right) + 2\frac{d\theta(x)}{dx}\left(\delta(x) - \delta(x - L_e)\right)\right).
\end{aligned}
$$

The strain rate damping induces a force density that is similarly evaluated as

$$\frac{\partial^2 M_d(x,t)}{\partial x^2} = c_s I(x)\frac{\partial^5 w_{rel}(x,t)}{\partial x^4 \partial t} + c_s \frac{d^2 I(x)}{dx^2}\frac{\partial^3 w_{rel}(x,t)}{\partial x^2 \partial t} + 2c_s \frac{dI(x)}{dx}\frac{\partial^4 w_{rel}(x,t)}{\partial x^3 \partial t}. \tag{6.4}$$

The width of a triangular beam $b(x)$ varies linearly with x as $b(x) = b_0 + \frac{x}{L}(b_L - b_0)$. Defining α, the tapering factor as

$$\alpha = \frac{b_0 - b_L}{b_0},$$

we can rewrite the width of the beam as

$$b(x) = b_0(1 - \alpha\frac{x}{L}),$$

which leads to a linear profile for various beam properties as the second moment of inertia, $I(x) = I_0(1 - \alpha\frac{x}{L})$; the flexural rigidity, $YI(x) = YI_0(1 - \alpha\frac{x}{L})$; and the mass per unit length $m(x) = m_0(1 - \alpha\frac{x}{L})$. Also, the piezoelectric coupling term $\theta(x)$ can be written as $\theta(x) = \theta_0(1 - \alpha\frac{x}{L})$, where $\theta_0 = -Y_p d_{31} b_0 z_{pc}$. The first derivatives of these *linear* terms reduce to a constant and their higher derivatives vanish and the force density terms in Eqs. (6.3)–(6.4) are reduced to

$$\begin{aligned}
\frac{\partial^2 M_m(x,t)}{\partial x^2} &= YI_0(1 - \alpha\frac{x}{L})\frac{\partial^4 w_{rel}(x,t)}{\partial x^4} - 2YI_0\frac{\alpha}{L}\frac{\partial^3 w_{rel}(x,t)}{\partial x^3},\\
\frac{\partial^2 M_d(x,t)}{\partial x^2} &= c_s I_0(1 - \alpha\frac{x}{L})\frac{\partial^5 w_{rel}(x,t)}{\partial x^4 \partial t} - 2c_s I_0\frac{\alpha}{L}\frac{\partial^4 w_{rel}(x,t)}{\partial x^3 \partial t}\\
\frac{\partial^2 M_e(x,t)}{\partial x^2} &= v(t)\left(\theta_0(1 - \alpha\frac{x}{L})\left(\frac{d\delta(x)}{dx} - \frac{d\delta(x - L_e)}{dx}\right)\right.\\
&\quad \left. -2\theta_0\frac{\alpha}{L}\left(\delta(x) - \delta(x - L_e)\right)\right),
\end{aligned}$$

respectively. Substituting the net force density $\frac{\partial^2 M(x,t)}{\partial x^2} = \frac{\partial^2 M_m(x,t)}{\partial x^2} + \frac{\partial^2 M_d(x,t)}{\partial x^2} + \frac{\partial^2 M_e(x,t)}{\partial x^2}$ into the structural governing equation (Eq. 6.1) and rearranging, we get

$$\begin{aligned}
&YI(x)\frac{\partial^4 w_{rel}(x,t)}{\partial x^4} + m(x)\frac{\partial^2 w_{rel}(x,t)}{\partial t^2} - \alpha\frac{2YI_0}{L}\frac{\partial^3 w_{rel}(x,t)}{\partial x^3}\\
&+ c_s I(x)\frac{\partial^5 w_{rel}(x,t)}{\partial x^4 \partial t} - \alpha\frac{2c_s I_0}{L}\frac{\partial^4 w_{rel}(x,t)}{\partial x^3 \partial t} + c_a\frac{\partial w_{rel}(x,t)}{\partial t}
\end{aligned}$$

(continued)

$$+\theta(x)\left(\frac{d\delta(x)}{dx}-\frac{d\delta(x-L_e)}{dx}\right)v(t)-\frac{2\alpha\theta_0}{L}\left(\delta(x)-\delta(x-L_e)\right)v(t)$$
$$=-m(x)\frac{d^2w_b(x,t)}{dt^2}-c_a\frac{dw_b(x,t)}{dt}. \tag{6.5}$$

Choosing the tapering factor α as the taper control parameter, we notice that when $\alpha = 0$ Eq. (6.5) reduces to the rectangular beam case derived by Erturk et al. [6].

6.1.2 Modal Response with Electrical Coupling

The modal response equation, highlighted in Fig. 6.2 is a key equation that sets up the state-space system for a tapered beam energy harvester.

We first impose separation of variables x and t for a single mode r on the damped structural governing equation in Eq. (6.5) by substituting $w_{rel}(x,t) = \eta_r(t)\phi_r(x)$ to get

$$\left(YI(x)\frac{d^4\phi_r(x)}{dx^4}-\alpha\frac{2YI_0}{L}\frac{d^3\phi_r(x)}{dx^3}\right)\eta_r(t)+\frac{c_sI_0}{YI_0}\left(YI(x)\frac{d^4\phi_r(x)}{dx^4}\right.$$
$$\left.-\alpha\frac{2YI_0}{L}\frac{d^3\phi_r(x)}{dx^3}\right)\frac{d\eta_r(t)}{dt}+\theta(x)v(t)\left(\frac{d\delta(x)}{dx}-\frac{d\delta(x-L_e)}{dx}\right)$$

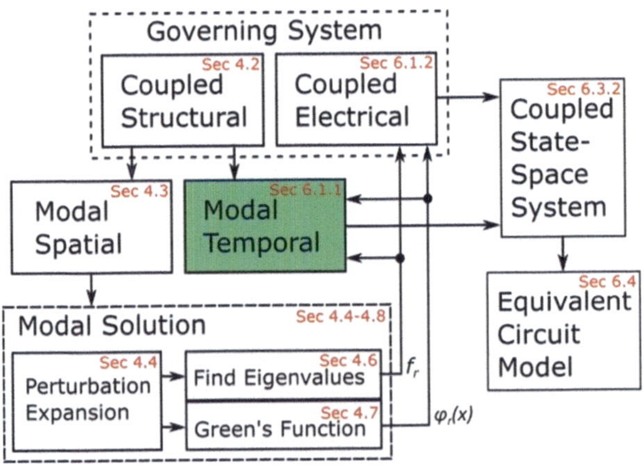

Fig. 6.2 Tapered PEH system formulation: temporal modal response equation

$$- 2\alpha\frac{\theta_0}{L}v(t)\left(\delta(x) - \delta(x - L_e)\right) + c_a\left(\phi_r(x)\frac{d\eta_r(t)}{dt} + \frac{dg(t)}{dt} + x\frac{dh(t)}{dt}\right)$$

$$+ m(x)\phi_r(x)\frac{d^2\eta_r(t)}{dt^2} + m(x)\frac{d^2g(t)}{dt^2} + m(x)\,x\frac{d^2h(t)}{dt^2} = 0. \tag{6.6}$$

The mode-shape equation is derived and its solution is presented in Chap. 4 where a perturbation expansion both in the eigenvalues and the mode shapes around the rectangular beam solution is applied to the mode-shape equation to solve the differential eigenvalue problem for the tapered beam.

The normalization conditions derived in Chap. 4 to force the effective mass to unity while ensuring orthogonality between the different mode shapes are

$$\int_0^L \phi_r(x)m(x)\phi_s(x)dx = \delta_{rs}, \text{ and}$$

$$\int_0^L \phi_s(x)\left(YI(x)\frac{d^4\phi_r(x)}{dx^4} - \frac{2}{L}\alpha YI_0\frac{d^3\phi_r(x)}{dx^3}\right)dx = \omega_r^2\delta_{rs}.$$

We multiply Eq. (6.6) by $\phi_s(x)$, integrate over x over the length of the beam, apply the normalization conditions above, and rearrange the differential equation in standard form to get

$$\frac{d^2\eta_r(t)}{dt^2}\delta_{rs} + 2\frac{c_sI_0\omega_r}{2YI_0}\omega_r\frac{d\eta_r(t)}{dt}\delta_{rs} + 2\frac{c_a\int_0^L\phi_s(x)m_0\phi_r(x)dx}{2m_0\omega_r}\omega_r\frac{d\eta_r(t)}{dt}$$

$$+ \eta_r(t)\omega_r^2\delta_{rs} + v(t)\left(\int_0^L\phi_s(x)\theta(x)\left(\frac{d\delta(x)}{dx} - \frac{d\delta(x-L_e)}{dx}\right)dx\right.$$

$$\left.-2\alpha\frac{\theta_0}{L}\int_0^L\phi_s(x)\left(\delta(x) - \delta(x-L_e)\right)dx\right) = -c_a\frac{dg(t)}{dt}\int_0^L\phi_s(x)dx$$

$$- c_a\frac{dh(t)}{dt}\int_0^L x\phi_s(x)dx - \frac{d^2g(t)}{dt^2}\int_0^L\phi_s(x)m(x)dx - \frac{d^2h(t)}{dt^2}\int_0^L x\phi_s(x)m(x)dx. \tag{6.7}$$

To simplify further, we use the sifting property of the derivatives of the Dirac delta function given as

$$\int_{-\infty}^{\infty}\frac{d^{(n)}\delta(x-x_0)}{dx^{(n)}}f(x)dx = (-1)^n\frac{df^{(n)}(x_0)}{dx^{(n)}}.$$

The first term for the $v(t)$ coefficient, which is the first part of the coupling factor χ_s, can be simplified as

$$\chi_{1s} = \int_0^L \phi_s(x)\theta(x)\left(\frac{d\delta(x)}{dx} - \frac{d\delta(x-L_e)}{dx}\right)dx$$
$$= \left(-\left.\frac{d\left(\phi_s(x)\theta(x)\right)}{dx}\right|_{x=0} + \left.\frac{d\left(\phi_s(x)\theta(x)\right)}{dx}\right|_{x=l_e}\right)$$

to

$$\chi_{1s} = -\left.\theta(x)\phi_s'(x)\right|_{x=0} + \left.\phi_s(x)\theta'(x)\right|_{x=0} + \left.\theta(x)\phi_s'(x)\right|_{x=l_e} + \left.\phi_s(x)\theta'(x)\right|_{x=l_e}.$$

We know from boundary conditions that $\phi_s'(0) = 0$ and $\phi_s(0) = 0$. Also we have $\theta(x) = -Y_p d_{31} z_{pc} b_0 (1 - \alpha\frac{x}{L})$. This means that $\theta(l_e) = -Y_p d_{31} z_{pc} b_0 (1 - \alpha\frac{L_e}{L})$ and $\theta'(l_e) = Y_p d_{31} z_{pc} b_0 \frac{\alpha}{L}$ to give

$$\chi_{1s} = -Y_p d_{31} z_{pc} b_0 (1 - \alpha\frac{L_e}{L})\phi_s'(l_e) + Y_p d_{31} z_{pc} b_0 \frac{\alpha}{L}\phi_s(l_e).$$

The second term for the $v(t)$ coefficient in Eq. (6.7), which is the second part of the coupling factor χ_s, can be simplified as

$$\chi_{2s} = -2\alpha\frac{\theta_0}{L}\int_0^L \phi_s(x)\left(\delta(x) - \delta(x-L_e)\right)dx = -2\alpha\frac{\theta_0}{L}\left(\phi_s(0) - \phi_s(l_e)\right).$$

We know that $\phi_s(0) = 0$ at the fixed end and that $\theta_0 = -Y_p d_{31} z_{pc} b_0$ to give

$$\chi_{2s} = -2Y_p d_{31} z_{pc} b_0 \frac{\alpha}{L}\phi_s(l_e).$$

The electrical-to-mechanical coupling factor in the modal dynamic equation is then $\chi_s = \chi_{1s} + \chi_{2s}$ to give

$$\frac{d^2\eta_r(t)}{dt^2}\delta_{rs} + 2\frac{c_s I_0 \omega_r}{2YI_0}\omega_r\frac{d\eta_r(t)}{dt}\delta_{rs} + 2\frac{c_a\int_0^L \phi_s(x)m_0\phi_r(x)dx}{2m_0\omega_r}\omega_r\frac{d\eta_r(t)}{dt}$$
$$+ \eta_r(t)\omega_r^2\delta_{rs} + \chi_s v(t) = -c_a\frac{dg(t)}{dt}\int_0^L \phi_s(x)dx$$
$$- c_a\frac{dh(t)}{dt}\int_0^L x\phi_s(x)dx - \frac{d^2g(t)}{dt^2}\int_0^L \phi_s(x)m(x)dx - \frac{d^2h(t)}{dt^2}\int_0^L x\phi_s(x)m(x)dx.$$

Evaluating for $s = r$ we get

$$\frac{d^2\eta_r(t)}{dt^2} + 2\frac{c_s I_0 \omega_r}{2YI_0}\omega_r \frac{d\eta_r(t)}{dt} + 2\frac{c_a \int_0^L \phi_r^2(x)dx}{2\omega_r}\omega_r \frac{d\eta_r(t)}{dt} + \eta_r(t)\omega_r^2 + \chi_r v(t)$$

$$= - c_a \frac{dg(t)}{dt} \int_0^L \phi_r(x)dx - c_a \frac{dh(t)}{dt} \int_0^L x\phi_r(x)dx$$

$$- \frac{d^2g(t)}{dt^2} \int_0^L \phi_r(x)m(x)dx - \frac{d^2h(t)}{dt^2} \int_0^L x\phi_r(x)m(x)dx.$$

Defining $\zeta_r = \frac{c_s I_0 \omega_r}{2YI_0} + \frac{c_a \int_0^L \phi_r^2(x)dx}{2\omega_r}$ and the excitation on the right-hand side as $F_r(t)$, we have the dynamic system given as

$$\frac{d^2\eta_r(t)}{dt^2} + 2\zeta_r\omega_r \frac{d\eta_r(t)}{dt} + \eta_r(t)\omega_r^2 + \chi_r v(t) = -F_r(t). \tag{6.8}$$

For ease of representation, we define the kth moment and the kth mass-weighted moment of a function $f(x)$ in Eqs. (6.9) and (6.10) as

$$P_k(f) = \int_0^L x^k f(x)dx, \text{ and} \tag{6.9}$$

$$P_k^{(m)}(f) = \int_0^L x^k f(x)m(x)dx, \text{ respectively .} \tag{6.10}$$

The damping ratio of the coupled dynamic system ζ_r includes components for viscous and strain rate damping and is given as

$$\zeta_r = \frac{c_s I_0 \omega_r}{2YI_0} + \frac{c_a P_0(\phi_r^2)}{2\omega_r}. \tag{6.11}$$

We can derive c_s and c_a if we have two measurements of the damping ratio, say ζ_j and ζ_k at two modes of vibrations, j and k, respectively. The damping coefficients

c_s and c_a can be derived for the tapered beam by simultaneously solving Eq. (6.11) at the two modes j and k, respectively [7]:

$$\begin{bmatrix} c_s \\ c_a \end{bmatrix} = \Omega_{jk} \begin{bmatrix} \frac{P_0(\phi_k^2)YI_0}{\omega_k I_0} & -\frac{P_0(\phi_j^2)YI_0}{\omega_j I_0} \\ -\omega_k & \omega_j \end{bmatrix} \begin{bmatrix} \zeta_j \\ \zeta_k \end{bmatrix}, \tag{6.12}$$

where $\Omega_{jk} = \frac{2\omega_j\omega_k}{P_0(\phi_k^2)\omega_j^2 - P_0(\phi_j^2)\omega_k^2}$. The cross-domain coupling factor χ_r for the rth mode is given as

$$\chi_r = \int_0^L \phi_r(x)\theta(x)\left(\frac{d\delta(x)}{dx} - \frac{d\delta(x-L_e)}{dx}\right)dx$$
$$- 2\alpha\frac{\theta_0}{L}\int_0^L \phi_r(x)\left(\delta(x) - \delta(x-L_e)\right)dx.$$

We use the sifting property of the derivatives of the Dirac delta function $\int_{-\infty}^{\infty} \frac{d^{(n)}\delta(x-x_0)}{dx^{(n)}} f(x)dx = (-1)^n \frac{df^{(n)}(x_0)}{dx^{(n)}}$ to simplify χ_r to get

$$\chi_r = \left[\frac{d\left(\phi_r(x)\theta(x)\right)}{dx}\right]_0^{L_e} - 2\alpha\frac{\theta_0}{L}\left(\phi_r(0) - \phi_r(L_e)\right)$$
$$= \left[\theta(x)\phi_r'(x) + \phi_r(x)\theta'(x)\right]_0^{L_e} - 2\alpha\frac{\theta_0}{L}\left(\phi_r(0) - \phi_r(L_e)\right).$$

We know from the boundary conditions that $\phi_r'(0) = 0$ and $\phi_r(0) = 0$. Also, using $\theta(0) = \theta_0 = -Y_p d_{31} z_{pc} b_0$, $\theta(L_e) = \theta_0(1 - \alpha\frac{L_e}{L})$, and $\theta'(L_e) = -\theta_0\frac{\alpha}{L}$ we get

$$\chi_r = \theta_0(1 - \alpha\frac{L_e}{L})\phi_r'(L_e) + \theta_0\frac{\alpha}{L}\phi_r(L_e). \tag{6.13}$$

The excitation term on the right-hand side of Eq. (6.8), $F_r(t)$, is evaluated as $F_r(t) = F_r^m(t) + F_r^c(t)$ where $F_r^m(t) = P_0^{(m)}\left(\phi_r(x)\frac{d^2w_b(x,t)}{dt^2}\right)$ is the actuation excitation term and $F_r^c(t) = c_a P_0\left(\phi_r(x)\frac{d^2w_b(x,t)}{dt^2}\right)$ is the damping-induced excitation term. Using $w_b(x,t) = g(t) + xh(t)$ for a cantilever beam, the excitation can be written as

$$F_r^m(t) = P_0^{(m)}(\phi_r)\frac{d^2g(t)}{dt^2} + P_1^{(m)}(\phi_r)\frac{d^2h(t)}{dt^2}, \text{ and}$$
$$F_r^c(t) = c_a P_0(\phi_r)\frac{dg(t)}{dt} + c_a P_1(\phi_r)\frac{dh(t)}{dt}. \tag{6.14}$$

When the rotational base excitation and the device body damping can be ignored, as is usually the case for energy harvesting devices, $F_r(t)$ can be reduced to $F_r(t) = P_0^{(m)}(\phi_r)\frac{d^2g(t)}{dt^2}$.

6.1.3 Electrical Circuit Equation with Mechanical Coupling

We next derive the electrical circuit equation highlighted in Fig. 6.3 to complete the electromechanical model of the tapered piezoelectric energy harvester. To derive the electrical circuit equation with mechanical coupling, we begin by considering the piezoelectric constitutive relation:

$$D_3(x,t) = d_{31}T_1 + \varepsilon_{33}^T E_3(x,t), \tag{6.15}$$

where $D_3(x,t)$ is the electric displacement, ε_{33}^T is the piezoelectric layers' permittivity at constant stress, and $E_3(x,t)$ is the electric field across the PZT layer. We substitute $T_1(x,t)$ in terms of $S_1(x,t)$ using the relation $T_1(x,t) = Y_pS_1(x,t) - d_{31}Y_pE_3(x,t)$ where Y_p is Young's modulus of the PZT layer. We then replace the permittivity at constant strain as $\varepsilon_{33}^T = \varepsilon_{33}^S + d_{31}^2 Y_p$ in Eq. (6.15). Moreover, we also write the PZT electric field in terms of the voltage across the PZT as

$$D_3(x,t) = d_{31}Y_pS_1(x,t) - \varepsilon_{33}^S\frac{v(t)}{h_p}. \tag{6.16}$$

The average bending strain in the piezoelectric layer can be obtained by integrating the bending strain expression in Eq. (4.3) over the piezoelectric layer thickness as

$$S_1(x,t) = -z_{pc}\frac{\partial^2 w_{rel}(x,t)}{\partial x^2}.$$

Substituting $S_1(x,t)$ in Eq. (6.16) we get

$$D_3(x,t) = -d_{31}Y_pz_{pc}\frac{\partial^2 w_{rel}(x,t)}{\partial x^2} - \varepsilon_{33}^S\frac{v(t)}{h_p}. \tag{6.17}$$

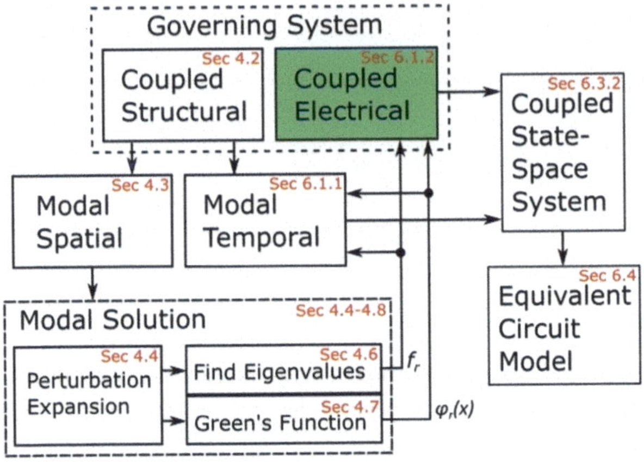

Fig. 6.3 Tapered PEH system formulation: coupled electric circuit equation

Applying Gauss's law on Eq. (6.17), the electric charge collected as a result of this electric displacement can be computed as the xy area integral over the device as

$$q(t) = \int_A \boldsymbol{D} \cdot \boldsymbol{n} dA = -\int_0^{L_e} \left(d_{31} Y_p z_{pc} b(x) \frac{\partial^2 w_{rel}(x,t)}{\partial x^2} + \varepsilon_{33}^S b(x) \frac{v(t)}{h_p} \right) dx,$$

where $\boldsymbol{D}$ is the electric displacement vector and $\boldsymbol{n}$ is the outward normal out the piezoelectric layer surface in the z-direction. Substituting $b(x) = b_0(1 - \alpha \frac{x}{L})$ we get

$$q(t) = -\int_0^{L_e} \left(d_{31} Y_p z_{pc} b_0 (1 - \alpha \frac{x}{L}) \frac{\partial^2 w_{rel}(x,t)}{\partial x^2} + \varepsilon_{33}^S b_0 (1 - \alpha \frac{x}{L}) \frac{v(t)}{h_p} \right) dx.$$

Taking the time derivative of the charge collected, we get the electric current as

$$i(t) = i_{mot}(t) - C_p \frac{dv(t)}{dt},$$

where the first $i_{mot}(t)$ is the motional current generated by the beam vibration given as

$$i_{mot}(t) = -\int_0^{L_e} \left(d_{31} Y_p z_{pc} b_0 (1 - \alpha \frac{x}{L}) \frac{\partial^3 w_{rel}(x,t)}{\partial x^2 \partial t} \right) dx,$$

while the second term is the current resulting from the voltage $v(t)$ generated across the piezoelectric layer with electric capacitance C_p given as

$$C_p = \frac{\varepsilon_{33}^S b_0 L_e}{h_p}\left(1 - \frac{\alpha}{2}\right). \tag{6.18}$$

C_p can also be calculated as $C_p = \frac{\varepsilon_{33}^S A_p}{h_p}$ where $A_p = L_e \frac{b_0 + b_L}{2} = L_e \frac{b_0 + b_0(1-\alpha)}{2} = L_e b_0(1 - \frac{\alpha}{2})$. Applying a load R_l across the energy harvester terminals, we get the equation for the load voltage $v(t)$ as

$$v(t) = -R_l i_{mot}(t) - R_l C_p \frac{dv(t)}{dt}.$$

Rearranging we get the nodal current balancing equation which can be expressed in the modal coordinates for all the modes present in the system using $w_{rel}(x,t) = \sum_{r=1}^{\infty} \eta_r(t)\phi_r(x)$ as

$$C_p \frac{dv(t)}{dt} + \frac{1}{R_l} v(t) = \sum_{r=1}^{\infty} \varphi_r \frac{d\eta_r(t)}{dt}, \tag{6.19}$$

where $\varphi_r = -d_{31} Y_p z_{pc} b_0 \int_0^{L_e} \left((1 - \alpha\frac{x}{L})\frac{\partial^2 \phi_r(x)}{\partial x^2}\right) dx$, which can further be simplified to

$$\varphi_r = -Y_p d_{31} z_{pc} b_0 \left((1 - \alpha\frac{L_e}{L})\phi_r'(L) + \frac{\alpha}{L}\phi_r(L)\right). \tag{6.20}$$

Note that both the cross-coupling factors φ_r and χ_r are equal when $L_e = L$ where the electrode spans the entire beam.

6.2 Steady-State Analysis

6.2.1 Steady-State Solution with Harmonic Inputs

To discover the closed-form steady-state solution of the modal equation of motion given in Eq. (6.8), we assume the translation and rotation of the device to be harmonic in time. This sets up the inputs $g(t)$ and $h(t)$ as

$$g(t) = g_0 e^{j\omega t}, \text{ and}$$
$$h(t) = h_0 e^{j\omega t},$$

where $j = \sqrt{-1}$, g_0 is the amplitude of the base translation, h_0 is the amplitude of the rotation angle, and ω is the excitation frequency. Knowing that the electromechanical system is linear, we expect the voltage response to also follow the harmonic profile of the input in the form $v(t) = V_0 e^{j\omega t}$ where V_0 is the amplitude of the harmonic voltage across the load. With the harmonic inputs and also ignoring the influence of damping to the device actuation, the excitation term $F_r(t)$ on the right-hand side of the modal equation of motion from Eq. (6.8) can be reduced to

$$F_r(t) = -\omega^2 \left(P_0^m(\phi_r) g_0 + P_1^m(\phi_r) h_0\right) e^{j\omega t}.$$

Imposing the same harmonic condition on the modal response, the modal equation of motion can be written as

$$\omega^2 \eta_r(t) + 2\zeta_r \omega_r \omega \eta_r(t) + \omega_r^2 \eta_r(t) = \left(\omega^2 \left(P_0^m(\phi_r) g_0 + P_1^m(\phi_r) h_0\right) - \chi_r V_0\right) e^{j\omega t}.$$

The steady-state solution of the modal equation of motion is then evaluated as

$$\eta_r(t) = \frac{\omega^2 \left(P_0^m(\phi_r) g_0 + P_1^m(\phi_r) h_0\right) - \chi_r V_0}{\omega_r^2 - \omega^2 + j2\zeta_r \omega_r \omega} e^{j\omega t}. \tag{6.21}$$

Similarly, the electrical circuit equation reduces to its steady-state formulation as

$$\left(\frac{1 + j\omega\tau_c}{R_l}\right) V_0 e^{j\omega t} = \sum_{r=1}^{\infty} \varphi_r \frac{d\eta_r(t)}{dt}, \tag{6.22}$$

where $\tau_c = R_l C_p$. Substituting $\eta_r(t)$ from Eq. (6.21) in Eq. (6.22) we get

$$\left(\frac{1 + j\omega\tau_c}{R_l}\right) V_0 = \sum_{r=1}^{\infty} \varphi_r \frac{j\omega \left(\omega^2 \left(P_0^m(\phi_r) g_0 + P_1^m(\phi_r) h_0\right) - \chi_r V_0\right)}{\omega_r^2 - \omega^2 + j2\zeta_r \omega_r \omega}.$$

Solving for V_0 we get

$$V_0 = \frac{\omega^2 \sum_{r=1}^{\infty} \frac{j\omega\varphi_r \left(P_0^m(\phi_r) g_0 + P_1^m(\phi_r) h_0\right)}{\omega_r^2 - \omega^2 + j2\zeta_r \omega_r \omega}}{\sum_{r=1}^{\infty} \frac{j\omega \chi_r \varphi_r}{\omega_r^2 - \omega^2 + j2\zeta_r \omega_r \omega} + \frac{1 + j\omega\tau_c}{R_l}}. \tag{6.23}$$

The voltage response $v(t)$ across a resistive load R_l due to harmonic base excitation of $w_b(x,t) = (g_0 + xh_0)\,e^{j\omega t}$ can be expressed as

$$v(t) = V_0 e^{j\omega t}. \tag{6.24}$$

Note that this relationship holds just as well if the resistive load in this formulation is replaced by the impedance of a reactive load Z_l for as long as the linearity of the system is preserved. Also note that the current through the load $i(t)$ and the instantaneous power output $P(t)$ can be derived from Eq. (6.24) as $i(t) = v(t)/R_l$ and $P(t) = v^2(t)/R_l$. We can also evaluate the instantaneous steady-state modal response of the beam by substituting V_0 from Eq. (6.23) in Eq. (6.21) to get

$$\eta_r(t) = \frac{1}{\omega_r^2 - \omega^2 + j2\zeta_r\omega_r\omega}\left(\omega^2\left(P_0^m(\phi_r)g_0 + P_1^m(\phi_r)h_0\right)\right.$$
$$\left. - \chi_r \frac{\omega^2 \sum_{r=1}^{\infty} \frac{j\omega\varphi_r\left(P_0^m(\phi_r)g_0 + P_1^m(\phi_r)h_0\right)}{\omega_r^2-\omega^2+j2\zeta_r\omega_r\omega}}{\sum_{r=1}^{\infty} \frac{j\omega\chi_r\varphi_r}{\omega_r^2-\omega^2+j2\zeta_r\omega_r\omega} + \frac{1+j\omega\tau_c}{R_l}}\right)e^{j\omega t},$$

and the instantaneous beam deflection relative to the base is given as

$$w_{rel}(x,t) = \sum_{r=1}^{\infty} \frac{\phi_r(x)}{\omega_r^2 - \omega^2 + j2\zeta_r\omega_r\omega}\left(\omega^2\left(P_0^m(\phi_r)g_0 + P_1^m(\phi_r)h_0\right)\right.$$
$$\left. - \chi_r \frac{\omega^2 \sum_{r=1}^{\infty} \frac{j\omega\varphi_r\left(P_0^m(\phi_r)g_0 + P_1^m(\phi_r)h_0\right)}{\omega_r^2-\omega^2+j2\zeta_r\omega_r\omega}}{\sum_{r=1}^{\infty} \frac{j\omega\chi_r\varphi_r}{\omega_r^2-\omega^2+j2\zeta_r\omega_r\omega} + \frac{1+j\omega\tau_c}{R_l}}\right)e^{j\omega t}. \tag{6.25}$$

6.2.2 *Numerical Evaluation of the Steady-State Solution*

The numerical evaluation of the steady-state solution of $v(t)$ and $w_{rel}(x,t)$ is a straightforward evaluation of Eqs. (6.24) and (6.25), respectively, after having computed the mode shapes $\phi_r(x)$ and resonance frequencies ω_r using Algorithm 4.1. The final expressions for $v(t)$ and $w_{rel}(x,t)$ suggest that the argument of each of the summations can be computed for each mode independently and summed up at the end before the steady-state responses are evaluated. Algorithm 6.1 describes the strategy for evaluating the steady-state response of a tapered piezoelectric energy harvester.

Algorithm 6.1 Steady-state analysis of a tapered beam energy harvester

procedure SteadyStateAnalysisOfTaperedBeam(r, N, N_m, x, t, g_0, h_0, ω, R_l, Q)

r: Mode number, N: Perturbation order, N_m: Number of modes,

$x = \{x_i\}\forall i = 1\cdots N_x,\ s.t.\, x_i \in (0, l)$

$t = 0\cdots t_{N_t}$

$\omega = \{\omega_i\},\ s.t.\, \omega_i$ is in the neighborhood of expω_r

g_0: Translational excitation amplitude

h_0: Rotational excitation amplitude

c_s and c_a are derived from expected Q-factor

Define $P_k(\phi_r) = \int_0^L x^k \phi_r(x)dx$

Define $P_k^m(\phi_r) = \int_0^L x^k \phi_r(x)m(x)dx$

Setup

Compute α ▷ Taper factor: $\alpha = \frac{b_0 - b_L}{b_0}$

$C_p = \frac{\varepsilon_{33}^S b_0 l}{h_p}(1 - \frac{\alpha}{2})$

$\tau_c = R_l C_p$

Modal Analysis

for $r \leftarrow 1, \max(2, N_m)$ **do**

Compute ω_r and $\phi_r(x)$ ▷ Call Algorithm 4.1

end for

Damping

$c_s = \frac{2\omega_1\omega_2}{\omega_1^2 - \omega_2^2}\frac{YI_0}{I_0}\left(\frac{\zeta_1}{\omega_2} - \frac{\zeta_2}{\omega_1}\right)$

$c_a = \frac{2\omega_1\omega_2}{\omega_1^2 - \omega_2^2}\left(\frac{\omega_2\zeta_1}{P_0(\phi_2^2)} - \frac{\omega_1\zeta_2}{P_0(\phi_1^2)}\right)$

Voltage estimate, $v(t)$

for $r \leftarrow 1, N_m$ **do**

Compute $\phi_r(l)$ and $\phi_r'(l)$

<u>Forcing function, F_r</u>

$F_r = -\omega^2\left(P_0^m(\phi_r)g_0 + P_1^m(\phi_r)h_0\right)$

<u>Parameters, ζ_r, χ_r</u>

$\zeta_r = \frac{c_s I_0 \omega_r}{2YI_0} + \frac{c_a P_0(\phi_r^2)}{2\omega_r}$

$\chi_r = -Y_p d_{31} z_{pc} b_0 \left((1-\alpha)\phi_r'(l) + \frac{\alpha}{L}\phi_r(l)\right)$

<u>Voltage magnitude contributions per mode</u>

$V_{0,r}^{num} = \frac{-j\omega\chi_r F_r}{\omega_r^2 - \omega^2 + j2\zeta_r\omega_r\omega}$

$V_{0,r}^{den,1} = \frac{j\omega\chi_r^2}{\omega_r^2 - \omega^2 + j2\zeta_r\omega_r\omega}$

end for

$V_0 = \frac{\sum_{r=1}^{\infty} V_{0,r}^{num}}{\sum_{r=1}^{\infty} V_{0,r}^{den,1} + \frac{1 + j\omega\tau_c}{R_l}}$

Return $v(t) = V_0 e^{j\omega t}$

Deflection estimate, $w_{rel}(x, t)$

for $r \leftarrow 1, N_m$ **do**

$W_{0,r}(x) = \frac{\phi_r(x)}{\omega_r^2 - \omega^2 + j2\zeta_r\omega_r\omega}(-F_r - \chi_r V_0)$

$W_0(x) = \sum_{r=1}^{\infty} W_{0,r}(x)$

end for

Return $w_{rel}(x, t) = W_0(x)e^{j\omega t}$

end procedure

6.3 Transient Analysis

6.3.1 Implicit Transient Solution

The modal response in Eq. (6.8) can be expressed using the Duhamel integral as

$$\eta_r(t) = \frac{1}{\omega_{rd}} \int\limits_0^t (-F_r(\tau) - \chi_r v(\tau))\, e^{-\zeta_r \omega_r (t-\tau)} \sin \omega_{rd}(t-\tau)\, d\tau, \tag{6.26}$$

where $\omega_{rd} = \omega_r \sqrt{1-\zeta_r^2}$ is the damped natural frequency of the rth mode. Solving Eq. (6.19) using the integrating factor $e^{\frac{t}{\tau_c}}$ and setting the constant of integration to zero assuming that $v(0) = 0$, we get

$$v(t) = e^{-\frac{t}{\tau_c}} \sum_{r=1}^{\infty} \varphi_r \int\limits_0^t e^{\frac{\tau_1}{\tau_c}} \frac{d\eta_r(\tau_1)}{d\tau_1} d\tau_1. \tag{6.27}$$

Substituting $\eta_t(t)$ from Eq. (6.26) in Eq. (6.27) we get an implicit solution for $v(t)$ as

$$v(t) = -\, e^{-\frac{t}{\tau_c}} \sum_{r=1}^{\infty} \varphi_r \int\limits_0^t e^{\frac{\tau}{\tau_c}} \frac{d}{d\tau_1} \left(\frac{1}{\omega_{rd}} \int\limits_0^{\tau_1} \left(F_r(\tau) \right.\right.$$
$$\left.\left. + \chi_r v(\tau) \right) e^{-\zeta_r \omega_r (\tau_1 - \tau)} \sin \omega_{rd}(\tau_1 - \tau)\, d\tau \right) d\tau_1$$

Conversely, eliminating $v(t)$ we get the implicit solution for $\eta_r(t)$ as

$$\eta_r(t) = -\frac{1}{\omega_{rd}} \int\limits_0^t \left(F_r(\tau) + \chi_r e^{-\frac{\tau}{\tau_c}} \sum_{r=1}^{\infty} \varphi_r \int\limits_0^{\tau} e^{\frac{\tau_1}{\tau_c}} \frac{d\eta_r(\tau_1)}{d\tau_1} d\tau_1 \right)$$
$$\cdot\, e^{-\zeta_r \omega_r (t-\tau)} \sin \omega_{rd}(t-\tau)\, d\tau.$$

6.3.2 State-Space Model of the Electromechanical System

We will now turn our attention to the highlighted block in Fig. 6.4 and build a state-space representation of the electromechanical model of the TPEH system. The modal response equation (Eq. 6.8) and the circuit equation (Eq. 6.19) form

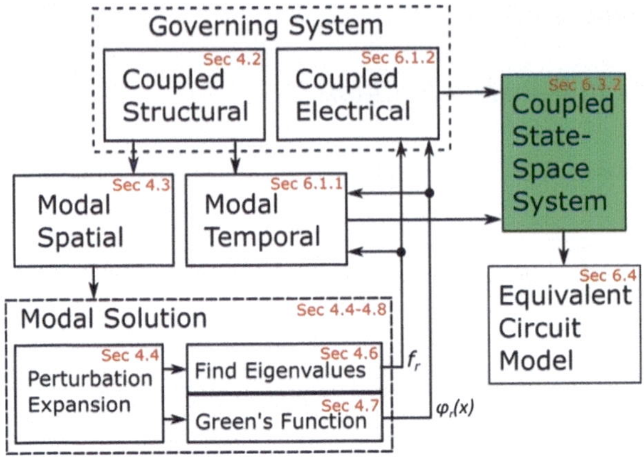

Fig. 6.4 Tapered PEH system formulation: coupled state-space system equation

the electromechanical system of equations as

$$\begin{aligned} &\ddot{\eta}_r(t) + 2\zeta_r\omega_r\dot{\eta}_r(t) + \eta_r(t)\omega_r^2 + \chi_r v(t) = -F_r(t), \text{ and} \\ &-\sum_{r=1}^{\infty} \chi_r\dot{\eta}_r + \frac{1}{R_l}v(t) + C_p\dot{v}(t) = 0, \end{aligned} \tag{6.28}$$

where $F_r(t)$, χ_r, and ζ_r are expressed in Eqs. (6.14), (6.13), and (6.11), respectively, and C_p and φ_r are expressed in Eqs. (6.18) and (6.20), respectively. We will now construct a first-order state-space model from these equations to be able to perform further analyses on the TPEH system including the transient analysis of the system using popular ODE solvers.

6.3.2.1 Single-Mode System

Notice that if the modes are well separated, then the summation in Eq. (6.28) can be dropped and the electromechanical system can be simplified to

$$\begin{aligned} &\ddot{\eta}_r(t) + 2\zeta_r\omega_r\dot{\eta}_r(t) + \eta_r(t)\omega_r^2 + \chi_r v(t) = -F_r(t), \text{ and} \\ &-\chi_r\dot{\eta}_r + \frac{1}{R_l}v(t) + C_p\dot{v}(t) = 0. \end{aligned}$$

We can set up the state derivative equation for this system using the state vector:

$$\mathbf{q}(t) = \begin{bmatrix} \dot{\eta}(t) \\ \eta(t) \\ v(t) \end{bmatrix}.$$

The state derivative equation for the rth mode can be written as

$$\frac{d\mathbf{q}(t)}{dt} = \boldsymbol{A}_r\mathbf{q}(t) + \boldsymbol{B}_r\mathbf{u}(t),$$

where

$$\mathbf{A}_r = \begin{bmatrix} -2\zeta_r\omega_r & -\omega_r^2 & \chi_r \\ 1 & 0 & 0 \\ \frac{\chi_r}{C_p} & 0 & -\frac{1}{R_l C_p} \end{bmatrix}.$$

Given that $F_r^m(t) = P_0^m(\phi_r)\frac{d^2g(t)}{dt^2} + P_1^m(\phi_r)\frac{d^2h(t)}{dt^2}$, we have

$$\mathbf{u}(t) = \begin{bmatrix} \ddot{g}(t) \\ \ddot{h}(t) \end{bmatrix},$$

and

$$\mathbf{B}_r = \begin{bmatrix} -P_0^m(\phi_r) & -P_1^m(\phi_r) \\ 0 & 0 \\ 0 & 0 \end{bmatrix}.$$

6.3.2.2 Multiple-Mode System

When the modes are not well separated or when a comprehensive multimodal model is nevertheless desired, we consider the first N_m modes of the system to set up the state-space model. The state vector should then include the mechanical modal responses and their derivatives for each of the modes as well as the output voltage as

$$\mathbf{q}(t) = \begin{bmatrix} \dot{\eta}_1(t) \\ \eta_1(t) \\ \dot{\eta}_2(t) \\ \eta_2(t) \\ \vdots \\ \dot{\eta}_{N_m}(t) \\ \eta_{N_m}(t) \\ v(t) \end{bmatrix}.$$

The input vector is the same as the single-mode case and the multimodal state derivative equation is given as

$$\frac{d\mathbf{q}(t)}{dt} = \boldsymbol{A}\mathbf{q}(t) + \boldsymbol{B}\mathbf{u}(t),$$

where the matrices $\boldsymbol{A}$ and $\boldsymbol{B}$ are

$$\boldsymbol{A} = \begin{bmatrix} -2\zeta_1\omega_1 & -\omega_1^2 & 0 & 0 & \cdots & 0 & 0 & -\chi_1 \\ 1 & 0 & 0 & 0 & \cdots & 0 & 0 & 0 \\ 0 & 0 & -2\zeta_2\omega_2 & -\omega_2^2 & \cdots & 0 & 0 & -\chi_2 \\ 0 & 0 & 1 & 0 & \cdots & 0 & 0 & 0 \\ \vdots & \vdots & \vdots & \vdots & \vdots & \vdots & \vdots & \vdots \\ 0 & 0 & 0 & 0 & \cdots & 2\zeta_{N_m}\omega_{N_m} & -\omega_{N_m}^2 & -\chi_{N_m} \\ 0 & 0 & 0 & 0 & \cdots & 1 & 0 & 0 \\ \frac{\chi_1}{C_p} & 0 & \frac{\chi_2}{C_p} & 0 & \cdots & \frac{\chi_{N_m}}{C_p} & 0 & -\frac{1}{R_l C_p} \end{bmatrix},$$

and

$$\boldsymbol{B} = \begin{bmatrix} -P_0^m(\phi_1) & -P_1^m(\phi_1) \\ 0 & 0 \\ -P_0^m(\phi_2) & -P_1^m(\phi_2) \\ 0 & 0 \\ \vdots & \vdots \\ -P_0^m(\phi_1) & -P_1^m(\phi_1) \\ 0 & 0 \\ 0 & 0 \end{bmatrix}, \text{ respectively.}$$

If we are only interested in $\mathbf{y}(t) = \begin{bmatrix} \eta_1(t) \\ \eta_2(t) \\ \vdots \\ \eta_{N_m}(t) \\ v(t) \end{bmatrix}$, the output equation,

$$\mathbf{y}(t) = \mathbf{Cq}(t) + \mathbf{Du}(t),$$

is set up using $\mathbf{D} = \mathbf{0}$, and

$$\boldsymbol{C} = \begin{bmatrix} 0\,1\,0\,0 \cdots 0\,0\,0 \\ 0\,0\,0\,1 \cdots 0\,0\,0 \\ \vdots\;\vdots\;\vdots\;\vdots\;\;\vdots\;\;\vdots\;\vdots\;\vdots \\ 0\,0\,0\,0 \cdots 0\,1\,0 \\ 0\,0\,0\,0 \cdots 0\,0\,1 \end{bmatrix}.$$

If instead we are interested in obtaining $w_{rel}(x,t) = \sum_{r=1}^{\infty} \eta_r(t)\phi_r(x)$ at a specific point x along the beam, we can set up the output vector as $\mathbf{y}(x,t) = \begin{bmatrix} w_{rel}(x,t) \\ v(t) \end{bmatrix}$, and the **C** matrix is set up as

$$C(x) = \begin{bmatrix} 0 \ \phi_r(x) \ 0 \ \phi_r(x) \ \cdots \ 0 \ \phi_r(x) \ 0 \\ 0 \quad 0 \quad 0 \quad 0 \quad \cdots 0 \quad 0 \quad 1 \end{bmatrix}.$$

Once the state-space system is set up, we can use a commercially available ODE solver such as one of the solvers in ODE toolbox in MATLAB to solve the system.

6.4 Equivalent Circuit Model Extraction

Energy harvester devices do not, in practice, service only a simple resistive load, but are instead connected to a signal conditioning circuit that includes a rectifier, a regulator, and an energy storage module, followed by the power sink which will also likely be reactive. A system-level FEA or the analytical model suggested in Sect. 6.3.2 can only model a resistive load. In order to simulate the model with nonresistive loads and external circuitry, we will develop the equivalent circuit model for the device which can then be simulated in any circuit simulator. This step is the culmination of the TPEH modeling workflow, as shown in Fig. 6.5, in that it uses the semi-analytical TPEH model developed thus far and produces an ECM that can be used in a design module which is capable of producing the figures of merits required for design discovery.

Equivalent circuit representations of energy harvesters help look at the structural device model as an electric circuit thus allowing its co-simulation with the EH

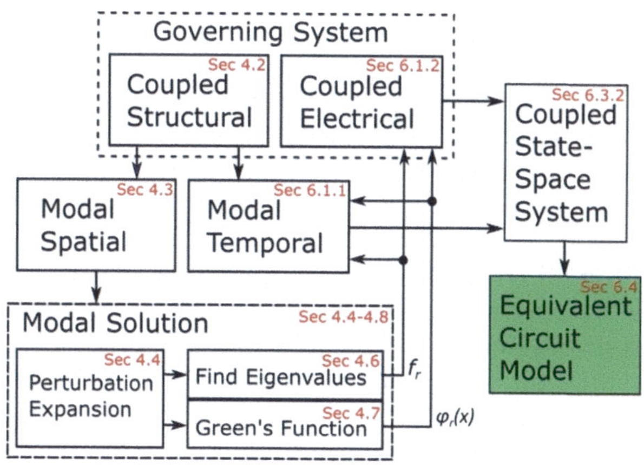

Fig. 6.5 Tapered PEH system formulation: equivalent circuit model

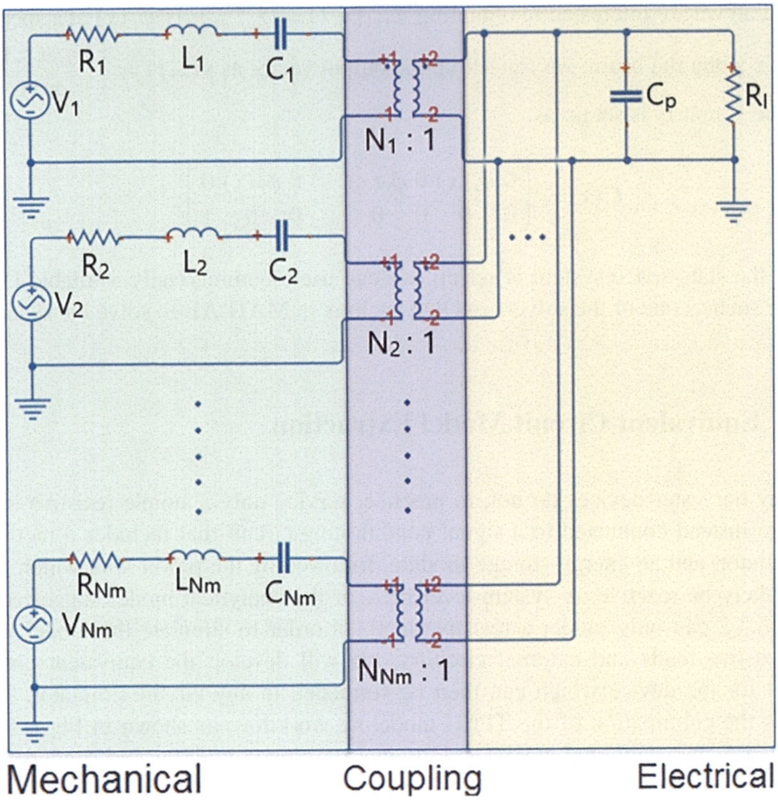

Fig. 6.6 Equivalent circuit model of the tapered energy harvester with multiple modes considered

acquisition and signal conditioning circuit using a circuit simulator. MEMS+ does provide a path to co-simulate the MEMS+ model of the device with a circuit in a Cadence simulation. This path, however, is computationally expensive in that a finite element solver is invoked for each iteration of the circuit simulation. Alternatively, the equivalent circuit model of the device integrates seamlessly in a circuit simulator with the EH circuit and simulates much faster. Such a simplified model is extremely useful when optimizing for a design in a loop over a large and multidimensional design space.

Elvin and Elvin [8] proposed piezoelectric generator equivalent circuit model based on a model constructed using the Rayleigh–Ritz approach as shown in the electromechanical formulations of Eqs. (6.8) and (6.19). The circuit network consists of branches for each of the vibrational modes, each composed of an inductor, a capacitor, a resistor, an ideal voltage source, and an ideal transformer as shown in Fig. 6.6. Note that the transformer models the coupling from the mechanical to the electrical side and vice versa. The resistor models the damping in

Table 6.1 Determining equivalent circuit parameters from electromechanical system model [9]

Equivalent circuit components at rth mode	Mechanical components
Charge: $q_r(t)$	Modal coordinate: $\eta_r(t)$
Current: $i_r(t)$	Modal velocity: $\dot{\eta}(t)$
Inductance: L_r	1 (mass-normalized)
Resistance: R_r	$2\zeta_r\omega_r$
Capacitance: C_r	$\frac{1}{\omega_r^2}$
Voltage source: $v(t)$	Forcing function:-$F_r(t)$
Ideal transformer ratio: N_r	Electromechanical coupling: χ_r

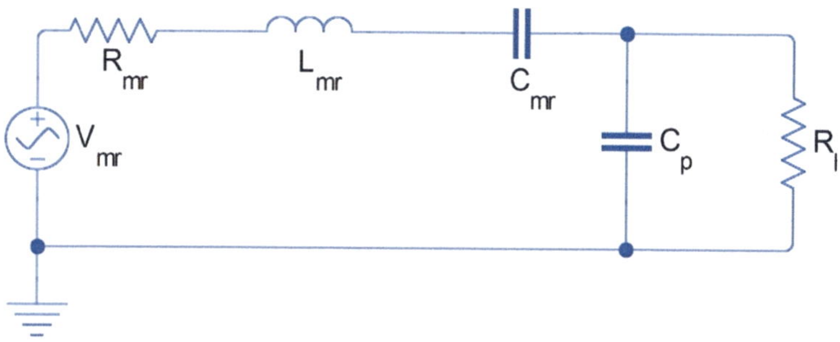

Fig. 6.7 Equivalent circuit model for each mode translated to the electric side

the system, the capacitor models the spring compliance, and the inductor models the mass which in this case is unity since the mode shapes have been mass-normalized.

The system simulation procedure includes first performing a theoretical modal analysis of the system to obtain natural frequencies and EH mode-shape functions which are then used to evaluate the ECM parameters R_r, L_r,C_r, F_r, and $V_r(t)$ using Table 6.1 before modeling the system in a circuit simulator for co-simulation with the electric EH circuit. The mechanical components in the table are identified from the results of modal analysis and by inspecting the modal response equation in Eq. (6.8).

One can move the equivalent circuit components in Fig. 6.6 to the electric side of the circuit using the relationships

$$L_{mr} = \frac{L_r}{N_r^2},$$

$$R_{mr} = \frac{R_r}{N_r^2},$$

$$C_{mr} = C_r N_r^2, \text{ and}$$

$$V_{mr} = \frac{V_r}{N_r}.$$

Each of the branches of the ECM in Fig. 6.6 can be moved to the electric side to form a single-mode electric side circuit shown in Fig. 6.7.

6.5 Equivalent Circuit Simulation

We performed AC analysis in three different ways. The first method uses the closed-form expression for AC analysis derived in Sect. 6.1. We also perform AC analysis on the ECM and compare the results with the closed-form results. To this end, we solved the circuit equations for the load voltage assuming harmonic acceleration input as described in Sect. 6.5.1. As a third option, we performed AC analysis using CppSim circuit analysis software. We used CppSim to perform the transient analysis of the ECM.

6.5.1 Analytical Solution for AC Analysis

To perform AC analysis of the ECM, we formulate the circuit equations using voltage loop analysis on each of the mechanical side branch and a nodal analysis on the electric side of the circuit as

$$V_r(t) = R_r i_r(t) + L_r \frac{di_r(t)}{dt} + \frac{1}{C_r} \int i_r(t)dt + N_r V_l(t).$$

This equation can be rewritten in terms of the charge $q_r(t)$ where $i_r(t) = \frac{dq_r(t)}{dt}$ as

$$V_r(t) = R_r \frac{dq_r(t)}{dt} + L_r \frac{d^2q_r(t)}{dt^2} + \frac{1}{C_r} q_r(t) + N_r V_l(t),$$

whereas the electric side nodal equation is set up as

$$C_p \frac{dV_l(t)}{dt} + \frac{V_l(t)}{R_p} + \frac{V_l(t)}{R_l} = \sum_{r=1}^{N_m} N_r i_r(t).$$

Considering harmonic inputs for AC analysis the above equations lead to the system of equations:

$$V_r(t) = \left(j\omega R_r - \omega^2 L_r + \frac{1}{C_r} \right) q_r(t) + N_r V_l(t), \text{ and}$$

$$\left(j\omega C_p + \frac{1}{R_p} + \frac{1}{R_l} \right) q_r(t) = j\omega \sum_{r=1}^{N_m} N_r q_r(t).$$

Eliminating q_r from these equations, we get the expression for V_l as

$$V_l(t) = \frac{\sum_{r=1}^{N_m} \frac{j\omega N_r V_r(t)}{j\omega R_r - \omega^2 L_r + \frac{1}{C_r}}}{\left(j\omega C_p + \frac{1}{R_p} + \frac{1}{R_l} \right) + \sum_{r=1}^{N_m} \frac{j\omega N_r^2}{j\omega R_r - \omega^2 L_r + \frac{1}{C_r}}}.$$

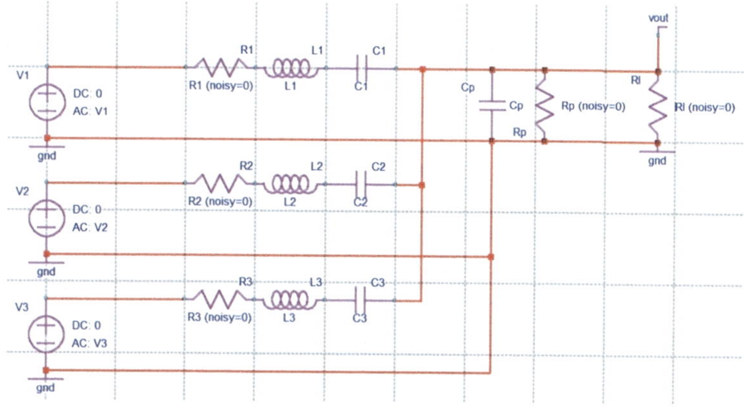

Fig. 6.8 ECM on the electric side implemented for 3 modes with CppSim

The impedance for optimal power based on impedance matching at the device interface is calculated using the electric side ECM in Fig. 6.7 and is given by

$$Z_{opt} = \frac{1}{j\omega_r C_p + \frac{N_r^2}{R_r + \frac{1}{j\omega_r C_r + j\omega_r L_r}}}.$$

If we are interested in resistive load, we take the magnitude of this optimal impedance to simulate the ECM at optimal output power. For reactive load, however, we can use Z_{opt} directly.

6.5.2 AC Analysis Using CppSim

We performed AC analysis sweep on the circuit using the CppSim circuit simulator for optimal loads calculated for each of the modes as well as for open-circuit voltage. The circuit we set up in CppSim is shown in Fig. 6.8.

6.5.3 Device Simulation Workflow Using ECM

Figure 6.9 shows the workflow we have devised for a tapered PEH simulation. We start with the device description that includes information from its layout and the technology used and compute the static capacitance of the piezoelectric layer. We then choose the number of perturbation terms N_p and use the perturbation-based tapered beam modal analysis algorithm presented in [10] to compute the N_m mode shapes $\{\phi_r\}_{r=1}^{N_m}$ and resonance frequencies $\{\omega_r\}_{r=1}^{N_m}$.

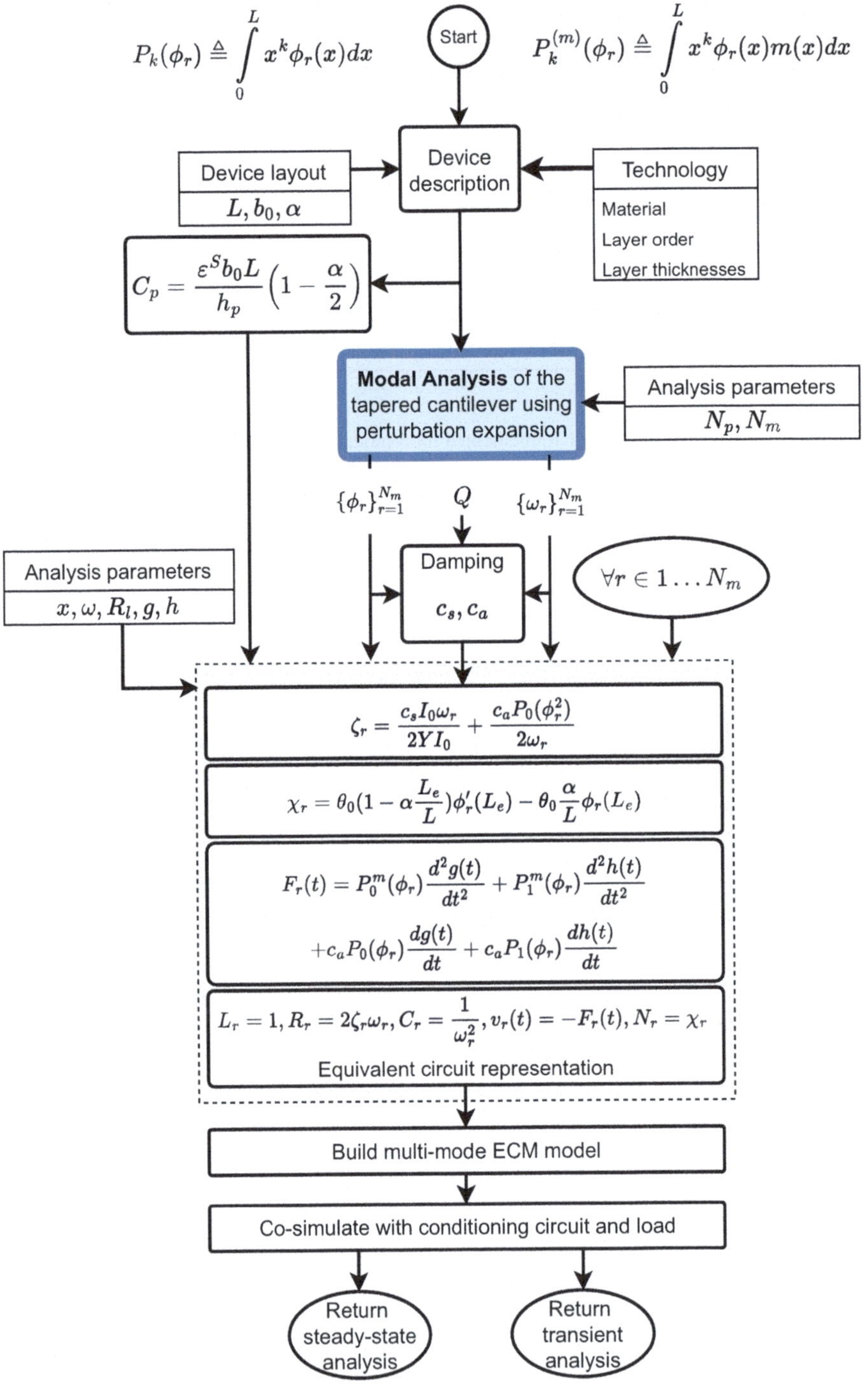

Fig. 6.9 Analysis flowchart for tapered PEH

We evaluate the damping coefficients using Eq. (6.12) for a specified Q-factor and then use them to evaluate the damping ratio ζ_r using Eq. (6.11). After evaluating the coupling factor χ_r using Eq. (6.13) and forcing function $F_r(t)$ using Eq. (6.14) for mechanical actuation $g(t)$ and $h(t)$, we evaluate the ECM parameters L_r, R_r, C_r, N_r, and $v_r(t)$ using relationships defined in Table 6.1. We construct a branch of the ECM for each of the modes and finally collect them together before using a circuit simulator to perform steady-state and transient analyses for the specified electric load R_l. The ECM extraction from FEM workflow has been validated against device measurement results through a variability analysis of ECM parameters in [11].

6.6 Simulation Results

To illustrate the accuracy of our model, we used the tapered beam structure used by Yang and Tang in [9]. The structure follows the geometry, material properties, and process stack used by Erturk in [12] and shown in Table 5.1 except that the cantilever is a tapered beam with base width $b_0 = 15$ mm and the tip width $b_L = 5$ mm.

6.6.1 *Model Validation for AC Analysis*

We first compare our ECM results with Erturk's [12] results for the straight cantilever PEH to establish the accuracy of our ECM machinery. The output power and voltage for optimum impedances at each of the modes are shown in Fig. 6.10. Additionally, we also present the results for output power at $R_l = 1$ MΩ for comparison with Erturk's results. We also show Erturk's results at the bottom of Fig. 6.10 to illustrate the correlation of the resonance peak locations and values between the two results.

We proceed to validate our ECM results against the tapered beam PEH results obtained by Yang and Tang [9], who themselves use Erturk's work as a benchmark for their FEM-based ECM and go on to present tapered PEH results using the same ECM. Again we present the output power and voltage for optimum impedances at each of the modes including also the outputs at $R_l = 1$ MΩ. The results are shown in Fig. 6.11 along with Yang and Tang's results at the bottom for comparison. Yang's results are obtained at a load of 1 MΩ and should be compared to the blue curve. Note that the resonance location and the magnitudes match well.

We used Yang and Tang's algorithm in [9] to build ECM using MEMS+ FEA simulations. Steady-state voltage and power outputs across the optimal load of about 1 MΩ were compared between the outputs from the ECM obtained from the proposed semi-analytical algorithm and Yang and Tang's FEA-based algorithm described in Sect. 3.4.4. The results shown in Fig. 6.12 indicate a reasonable match between the two when using only 5 perturbation terms.

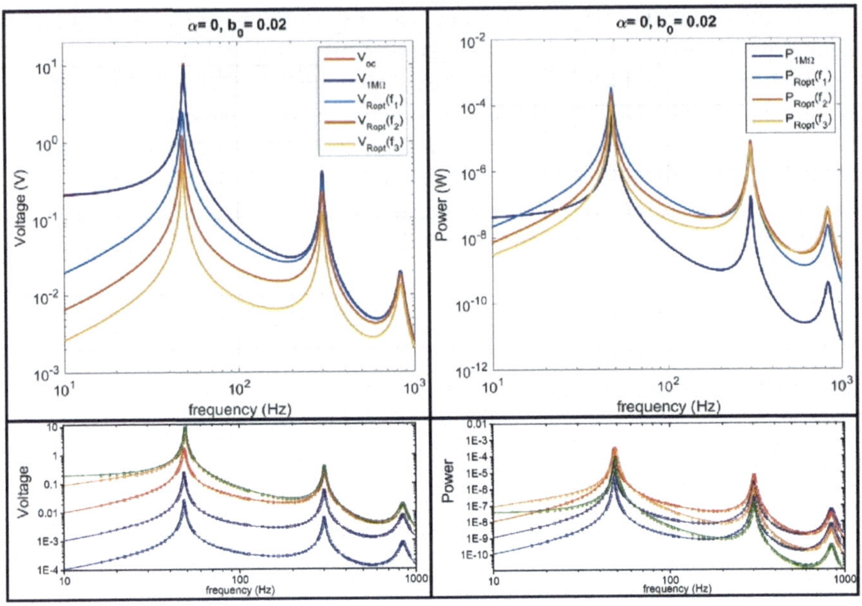

Fig. 6.10 ECM output comparison with Erturk's straight cantilever PEH [9]

6.6.2 *Beam Deflection AC Analysis*

Figure 6.13 shows the beam deflection magnitude estimated from AC analysis at different points along the length of the beam. Notice that the magnitude at the tip is consistently high around the resonance peaks. It does dip below the deflection magnitudes of points on the second half of the beam for a frequency range between the second and third modes. The low curvature close to the tip results in lower deflection amplitude for the three-quarter mark where the tip slope is high for the second mode.

6.6.3 *Transient Analysis*

The transient analysis of the TPEH is performed using the state-space model of the TPEH with solvers in MATLAB. The step response results for output voltage and the tip deflection of the tapered PEH used by Yang and Tang in [9] for a load resistance value of 1 MΩ are shown in Fig. 6.14. The step input is set up with a rise time of 1 μs and a fall time of 1 μs. The pulse width is kept large enough at 4s with a period of 5s to allow the response to settle. The 2% settling time for the tip deflection is at 0.812 s and that of the output voltage is 0.7 s. Notice that the output

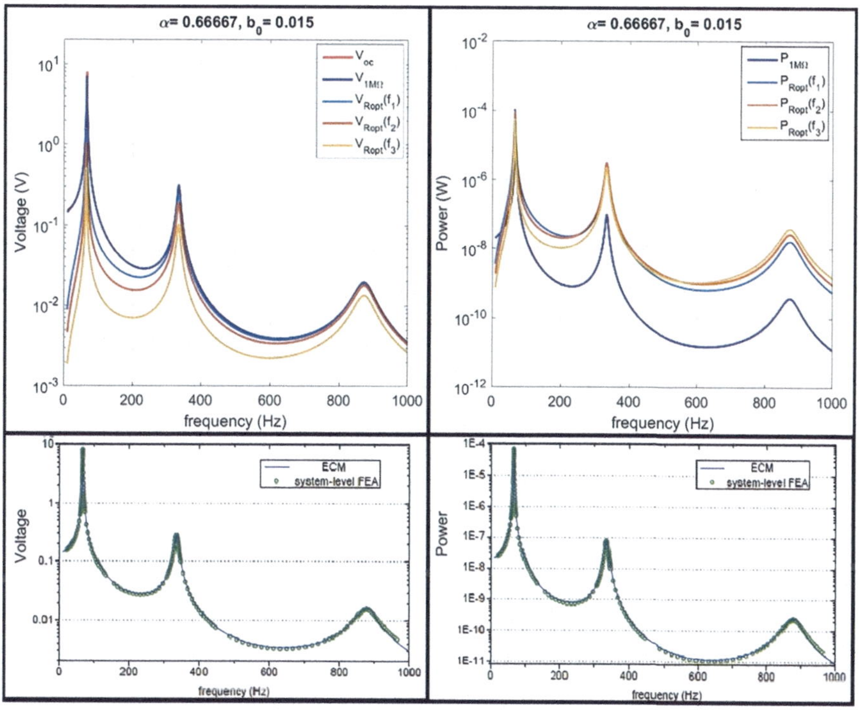

Fig. 6.11 ECM output comparison with Erturk's straight cantilever PEH [9]

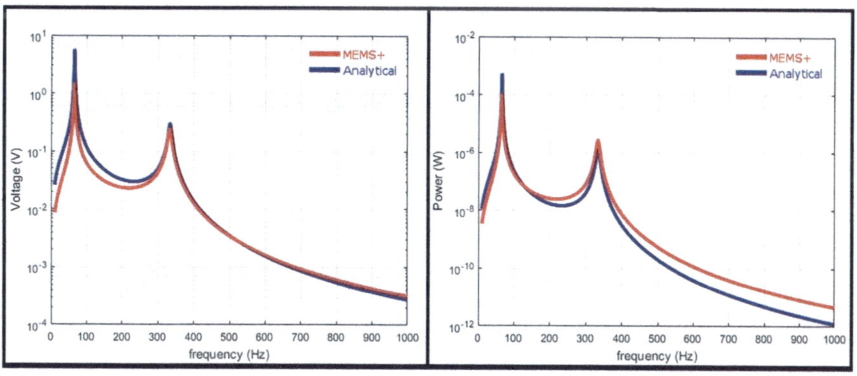

Fig. 6.12 Comparison of voltage and power across optimal load with ECM from Yang and Tang's FEA-based algorithm

voltage response to a step input settles to around 0 V, whereas the device structure settles in a bent state. The impulse response results for the output voltage and the tip deflection of the same TPEH are shown in Fig. 6.15.

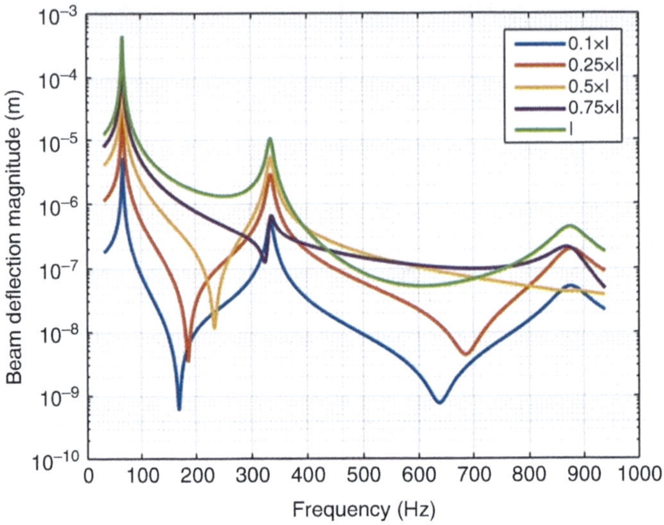

Fig. 6.13 Beam deflection magnitude at various points along the beam for Yang's tapered EH

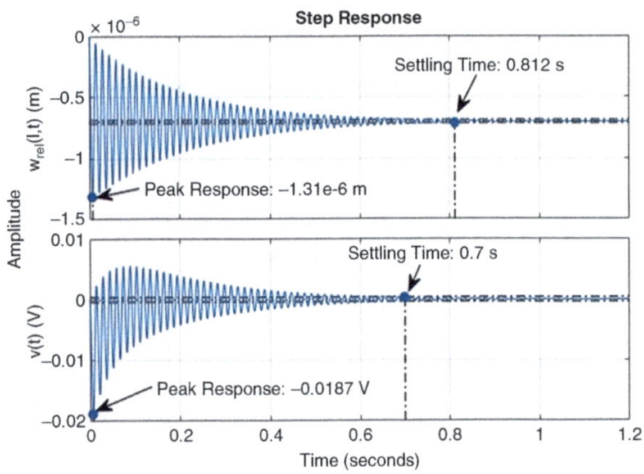

Fig. 6.14 Step response of Yang's tapered beam [9]

6.6.4 Timing Comparison with FEM

We perform a timing comparison for the model construction, modal analysis, ECM development, and simulation including AC and transient analyses between the proposed semi-analytical modal analysis-based algorithm implemented with varying numbers of perturbation terms and MEMS+ model with 6 spatial nodes. The results shown in Fig. 6.16 indicate more than $\times 150$ speedup when using the proposed algorithm with 5 perturbation terms and more than $\times 110$ speedup when using 20 perturbation terms.

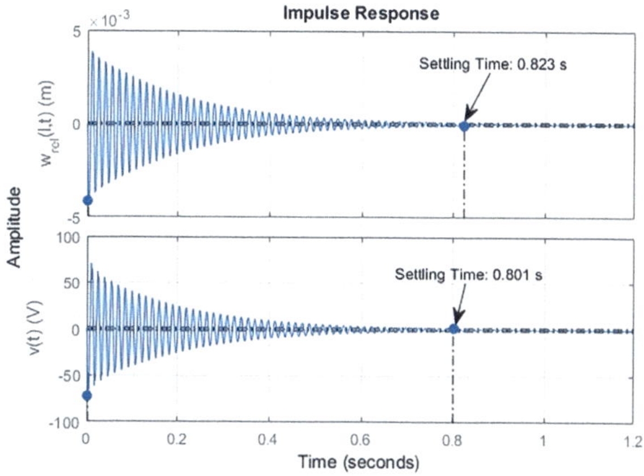

Fig. 6.15 Impulse response of Yang's tapered beam

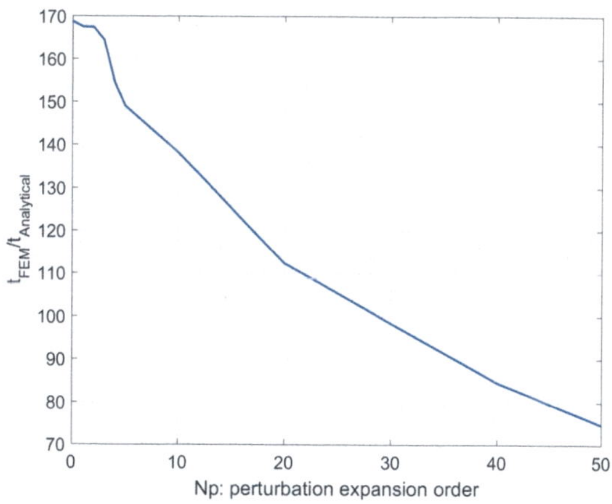

Fig. 6.16 Timing comparison between proposed algorithm and MEMS+

6.7 Conclusion

In this chapter, we show that a working complete electromechanical model of the tapered piezoelectric energy harvester can be developed based on the perturbation expansion-based modal analysis algorithm developed in Chap. 4. Once the expressions of the system parameters of the tapered PEH electromechanical system were found, the development of the state-space model and consequently the equivalent circuit model was straightforward. The closed-form expressions for voltage, power, and tip deflection AC analysis were formulated. Results were compared and cross-validated from a few different paths including the closed-form expressions, the state-space model, and the ECM via CppSim, a circuit simulator. The ECM itself was validated in a few different ways. The steady-state results were compared against published results and also against FEM-based ECM simulation. The proposed model has been shown to be very fast compared to FEM simulations and has great potential in enabling exhaustive design searches and uncertainty quantification strategies.

References

1. *CoventorWare Field Solver Reference*. Coventor Inc. 2016.
2. "Han-Chung" Wang. "Generalized hypergeometric function solutions on the transverse vibration of a class of nonuniform beams". In: *Journal of Applied Mechanics* (1967), pp. 702–708.
3. William Weaver Jr, Stephen P Timoshenko, and Donovan Harold Young. *Vibration problems in engineering*. John Wiley & Sons, 1990.
4. S Timoshenko and DH Young. "Vibration Problems in Engineering, (1955)". In: *Van Nostrana* (1968).
5. AH Meitzler et al. *IEEE standard on piezoelectricity*. 1988.
6. Alper Erturk and Daniel J Inman. "A distributed parameter electromechanical model for cantilevered piezoelectric energy harvesters". In: *Journal of vibration and acoustics* 130.4 (2008), p. 041002.
7. Alper Erturk and Daniel J Inman. "On mechanical modeling of cantilevered piezoelectric vibration energy harvesters". In: *Journal of Intelligent Material Systems and Structures* (2008).
8. Niell G Elvin and Alex A Elvin. "A general equivalent circuit model for piezoelectric generators". In: *Journal of Intelligent Material Systems and Structures* 20.1 (2009), pp. 3–9.
9. Yaowen Yang and Lihua Tang. "Equivalent circuit modeling of piezoelectric energy harvesters". In: *Journal of Intelligent Material Systems and Structures* (2009), pp. 2223–2235.
10. Wajih U. Syed and Ibrahim M. Elfadel. "Vibration Model of a Prismatic Multilayered Tapered Cantilever Using Perturbation Analysis". In: *Journal of Sound and Vibration* (2019).
11. Wajih Umar Syed, Alabi Bojesomo, and Ibrahim M Elfadel. "Variation-aware modelling of micro-scale piezoelectric energy harvesters". In: *2018 Symposium on Design, Test, Integration & Packaging of MEMS and MOEMS (DTIP)*. DTIP. 2018, pp. 1–4.
12. Alper Erturk and Daniel J Inman. "Issues in mathematical modeling of piezoelectric energy harvesters". In: *Smart Materials and Structures* 17.6 (2008), p. 065016.

Chapter 7
Uncertainty Quantification Using Monte Carlo Methods

In this chapter, we will discuss issues in the uncertainty quantification (UQ) of MEMS and the various strategies for quantifying and propagating uncertainties in MEMS models. We will focus on nonintrusive UQ strategies with special emphasis on Monte Carlo methods which we will use to generate UQ results for the tapered PEH model that we have developed in the previous chapters. We will also discuss stochastic collocation, a spectral method that is a nonintrusive fast alternative to Monte Carlo for uncertainty quantification.

7.1 Uncertainty Model of MEMS Process Variability

The spread of MEMS in diverse applications that operate under varying conditions and have varying degrees of device fidelity requirements has put increasing demands on designing reliable and robust MEMS devices. Accurate rapid computational models of these devices are required to plug into design optimization loops. The earlier chapters of this book build on past work on energy harvester analytical models to build a fast semi-analytical model for tapered beam piezoelectric energy harvesters. These deterministic simulation methods, however, do not take into account the uncertainties in the eventually fabricated device in its geometry and material properties that arise from the inherent variability in the MEMS fabrication process. The uncertainties are even greater in the case of low-cost manufacturing processes that result in residual stresses in the device structure, irregular surface topography, chemical contamination, imprecise depositions and etching, and side wall angles. In the case of a piezoelectric device, the foundry platform tolerance in the thickness of the piezoelectric layer is crucial to consider in the design process for two reasons. The piezoelectric material, such as AlN, is stiffer than the remaining stack, hence contributing significantly to the net structure stiffness and consequently influencing the resonance frequency and the general device performance. Secondly,

W. U. Syed, I. (Abe) M. Elfadel, *Tapered Beams in MEMS*,
https://doi.org/10.1007/978-3-031-66391-8_7

the transduction mechanism is also very sensitive to the piezoelectric layer thickness. For these reasons, in this work on uncertainty quantification, we have mainly considered the propagation of the piezoelectric layer thickness variability to the device performance metrics.

A good MEMS design strategy should produce devices that are robust enough to tolerate variability in material properties and fabrication processes both between different wafers and spatially across a single wafer. In this work, we apply different uncertainty quantification strategies on tapered beam piezoelectric energy harvesters to efficiently quantify the effect of process variability to help with EH device design. In doing so we leverage the fast semi-analytical model for tapered beam energy harvesters derived in Chaps. 4 and 6.

Two important challenges are faced when devising an uncertainty quantification strategy for MEMS. The first is to identify and, subsequently, model the uncertainty sources that affect the device's performance. The uncertainty sources considered in this work are either from variability in material properties or the foundry processes. In either case, the nature of uncertainty is hard to determine and is often approximated using a standard distribution [1]. The distribution function choice can be influenced by knowledge of the contribution of the uncertain parameter to the device physics within its operating range. Where detailed experimental data is available from the foundry data-driven probability distribution estimates can be computed and used in uncertainty quantification [2, 3]. The second challenge is to propagate the uncertainty model of the parameters through the system to model the uncertainty in the quantities of interest. Various intrusive and nonintrusive methods for uncertainty propagation are discussed in Sect. 7.2.2.

In this chapter, we explain the sources of variability in MEMS and their uncertainty models. We then compare the popular UQ strategies before elaborating on the Monte Carlo simulation. Finally, we present uncertainty quantification results using Monte Carlo simulation on the tapered beam EH model derived in previous chapters.

7.1.1 Sources of Process Variability

Variations are generally characterized as within-die (WID), die-to-die (D2D), wafer-to-wafer (W2W), and between manufacturing runs or lot-to-lot (L2L) [4, 5]. Process variations and the consequent uncertainty in device parameters can be systematic or random, or spatially or temporally correlated. Foundries often use a systematic process of characterizing test structures to discover these variation trends.

The variability in the fabricated structure is introduced as a result of tolerances in the execution of various MEMS processes such as PECVD, LRIE, isotropic etch, DRIE, and sputter metallization. Thermally induced on-wafer variations (D2D) are caused by wafer edge effects and apparatus inconsistencies. The wafer-to-wafer (W2W) variations depend upon how well conditions are stabilized between their use for different wafers. Photoresist chemistry in wet etching is difficult to control

Table 7.1 Some device uncertainties that impact the output

Parameter	Description	Variability source	Variability model source
Y_i	Elasticity modulus of layer i	Process variability, material inconsistencies	Deposition effects on material properties study [3]
ρ_i	Density of layer i	Process variability, material inconsistencies	Manufacturing study, material data
t_i	Thickness of layer i	Deposition tolerance	Foundry data
w_i, w_{pm}	Widths of layer i and proofmass	Etch tolerance	Foundry data
l_b, l_{pm}	Lengths of layer i and proofmass	Etch tolerance, design uncertainty	Foundry data, modeling error
θ_k	Sidewall angle of edge k	Etch rate tolerance	Foundry data
P	Cavity pressure	Getter function tolerance	Foundry data
μ	Fluid viscosity	Process variability	Foundry data

with precision and leads to variability in etching. Some variability sources were identified for the EH device in the sealed cavity and tabulated in Table 7.1.

7.1.2 *Developing Variability Models of Parameters with Aleatoric Uncertainty*

Uncertainty quantification in device performance requires accurate variability models of the system parameters. The variability model translates to discovering the probability density functions of the device-specific parameters. Where available, measurements on device characterization test structures help establish these distributions. In the absence of sufficient data, one can assume the PDF to be a standard distribution such as a Gaussian or a uniform distribution, and characterization experiments conducted at the foundry along with the fabrication history using the various processes can help establish the distribution parameters.

Nonparametric density estimation does not make assumptions about the nature of randomness but instead directly determines the uncertainty model from measured data. Researchers have used kernel density estimation (KDE) to estimate the PDF at any point from the sum of the kernel function contributions centered at the data points. The kernels help interpolate the density function in between data points.

For this work, we assume a normal distribution for our parameter of interest, namely, the piezoelectric layer thickness. We set the mean value of the distribution as the nominal design target thickness and the standard deviation as a third of the process tolerance for the feature.

7.2 Uncertainty Quantification Strategies for MEMS

Traditionally, the MEMS designers used rules of thumb to subjectively account for variability in the fabrication process and the consequent uncertainties in the fabricated devices through conservative designs keeping safety margins before the device performance falls below performance specifications. This leads to revenue loss and a lost opportunity for device optimization. Over time, as the awareness of the issues with process variability grew along with the demand for more reliable devices, designers have increasingly used Monte Carlo (MC) simulation-based strategies to propagate uncertainties in input parameters to device performance. Researchers have used commercial FEM software to simulate MEMS devices using Monte Carlo sampling to investigate the effect of the uncertainties in geometrical features of the actuator [6] along with air gap dimensions of devices [7]. Researchers have proposed some reliability-based design flows for different MEMS devices under uncertainties [5, 8–10] using MC simulations. While these MC-based strategies are fairly easy to implement, they are computationally expensive and can become impractical when the underlying MEMS device models comprise accurate mesh-based FEA models.

7.2.1 *Identifying Uncertainty in the Model*

Any UQ algorithm starts with understanding the aleatoric uncertainty embedded in the device model and characterizing this uncertainty in terms of estimated probability density functions for each of the parameters that exhibit such uncertainty and that have a significant impact on device output and performance. Before applying a UQ algorithm then, one must first prepare the statistical domain of the problem in the following way:

1. Identify the device design parameters that have a significant impact on the device performance and that are prone to uncertainties. Let these form the vector $\boldsymbol{\xi}$ of uncertain parameters.
2. Estimate the joint probability density function $p(\boldsymbol{\xi})$, for these parameters based on foundry data and any knowledge regarding the physics that impacts the parameter.

7.2.2 *Nonintrusive and Intrusive Uncertainty Quantification*

Nonintrusive or *sampling-based* solvers treat the underlying device model as a black box, in the way that they generate parameter samples and run the device model for each sample as one would simulate the model for a deterministic set of parameters, before using ensemble methods to evaluate desired device outputs and statistics. Methods using these solvers generally include the following steps [11, 12]:

1. For the given number of realizations N, generate N samples $\left\{\boldsymbol{\xi}^j\right\}_{j=1}^{N}$ for the random vector $\boldsymbol{\xi} \in \Omega$.
2. Solve the coupled computational model of the device $\mathcal{F}((\mathbf{x}(t, \boldsymbol{\xi}^j), \boldsymbol{\xi}^j) = 0$ for each sample $\boldsymbol{\xi}^j$ to generate states $\left\{\mathbf{x}(t, \boldsymbol{\xi}^j)\right\}_{j=1}^{N}$ and outputs $\left\{y(\boldsymbol{\xi}^j)\right\}_{j=1}^{N}$ for each realization.
3. Post process $\left\{\mathbf{x}(t, \boldsymbol{\xi}^j)\right\}_{j=1}^{N}$ and $\left\{y(\boldsymbol{\xi}^j)\right\}_{j=1}^{N}$ to obtain performance measures of the device such as the mean output power, resonance frequency spread, or the device yield.

Nonintrusive solvers are very easy to implement when the device model is readily available as a circuit simulation or an FEA model. Monte Carlo is by far the most popular nonintrusive solver class. Model fitting through regression of statistical data is also often used as another flavor of nonintrusive solvers. Stochastic collocation is a popular nonintrusive stochastic spectral method.

Intrusive solvers convert the stochastic system equation $\mathcal{F}((\mathbf{x}(t, \boldsymbol{\xi}), \boldsymbol{\xi}) = 0$ to a rather large deterministic system of equations which can be directly solved to obtain $\mathbf{x}(t, \boldsymbol{\xi})$. Though intrusive solvers are harder to set up in that they require modifying the system equation, thc benefit lies in the fact that the solution has to be evaluated only once instead of for each realization of the random variables as in the nonintrusive solvers. However, in the presence of a multitude of uncertain parameters, this one-time computation may turn out to be very expensive. The intrusive solvers generally follow the following flow [11]:

1. Identify K basis functions $\{\Psi_k(\boldsymbol{\xi})\}_{k=1}^{K}$ based on the uncertainty distribution modes.
2. Approximate $\mathbf{x}(t, \boldsymbol{\xi})$ as an analysis in $\left\{\hat{x}_k(t)\right\}_{k=1}^{K}$ in a subspace of these K basis functions as $\mathbf{x}(t, \boldsymbol{\xi}) \approx \tilde{\mathbf{x}}(t, \boldsymbol{\xi}) = \sum_{k=1}^{K} \hat{x}_k(t)\Psi_k(\boldsymbol{\xi})$.
3. Select a set of K testing functions $\{\Phi_k(\boldsymbol{\xi})\}_{k=1}^{K}$.
4. Set up a deterministic computational system enforcing the functional $\mathbb{F}((\mathbf{x}(t, \boldsymbol{\xi}^j), \boldsymbol{\xi}^j)$ to be orthogonal to each of the testing functions as

$$\left\langle \mathbb{F}((\mathbf{x}(t, \boldsymbol{\xi}^j), \boldsymbol{\xi}^j), \Phi_k(\boldsymbol{\xi}) \right\rangle_{\Omega, p(\boldsymbol{\xi})} = 0 \; for \; k = 1, \cdots, K..$$

5. Solve the above equation for the analysis components $\left\{\hat{x}_k(t)\right\}_{k=1}^{K}$ so as to evaluate the corresponding stochastic solution as $\tilde{\mathbf{x}}(t, \boldsymbol{\xi}) = \sum_{k=1}^{K} \hat{x}_k(t)\Psi_k(\boldsymbol{\xi})$.

7.3 Nonintrusive Uncertainty Quantification Strategies

We will discuss two popular nonintrusive UQ strategies used for MEMS, namely, Monte Carlo simulation and stochastic collocation simulation.

7.3.1 *Monte Carlo Simulation*

Monte Carlo (MC) methods present a widely used approach to propagate model and parameter uncertainty to evaluate output statistics. A standard technique to obtain random realizations of the input data is to generate a finite sequence of N random samples $\xi^1, \xi^2, \ldots, \xi^N$ by a *pseudo*-random number generator according to their joint probability density function $p(\boldsymbol{\xi})$. After assigning the random parameter values to the system equation, it is solved for a deterministic solution. Stochastic moments are estimated from the data assuming N is large enough. A histogram of output realizations helps estimate the probability density function of the output. Monte Carlo methods are robust since they do not impose any regularity requirements. Moreover, the methods are noninvasive and hence easy to implement. The downside, however, is that the convergence rate of MC methods is very low where the error is asymptotically proportional to $\frac{1}{\sqrt{N}}$ and hence more simulations are required to exploit the *law of large numbers* for more accuracy.

The Monte Carlo analysis of the tapered beam EH with regards to the quantifying uncertainties in the outputs of modal, steady-state, and transient analysis is described in the Algorithm 7.1 in the inset below.

Algorithm 7.1 Monte Carlo simulation-based UQ of a tapered PEH

procedure MCUQOFTPEH(*ModalAnalysis, SteadyStateAnalysis,* and *BuildECM*) *ModalAnalysis*: Algorithm 4.1, *SteadyStateAnalysis*: Algorithm 6.1

Generate $\Theta_N = \left\{\boldsymbol{\xi}^j\right\}_{j=1}^{N}$ for the uncertain parameters in $\boldsymbol{\xi} \in \Omega$

Invoke *ModalAnalysis*$\forall\boldsymbol{\xi}^j$

Compute $p(f_r)$, and $p(\bar{\kappa}_r)$ where $\bar{\kappa}_r$ is the mode-shape normalized average curvature

Invoke *SteadyStateAnalysis* $\forall\boldsymbol{\xi}^j$

Compute $p(P_{r,max})$, $p(V_{r,max})$, and $p(w_{\max})$

Invoke *BuildECM*$\forall\boldsymbol{\xi}^j$

Compute $p(C_p)$, $p(R_r)$, $p(C_r)$, $p(N_r)$, $p(fg_r)$

Compute PDFs of timing metrics

end procedure

7.3.2 Stochastic Collocation Simulation

Stochastic collocation is a stochastic spectral method that estimates the state vector $\mathbf{x}(t, \boldsymbol{\xi})$ and the output $y(\boldsymbol{\xi})$ as linear combinations of generalized polynomial chaos basis functions $\{H_k(\boldsymbol{\xi})\}_{k=1}^{K}$. Exploiting the orthonormality of $H_k(\boldsymbol{\xi})$, the projection of these vectors on the basis $H_k(\boldsymbol{\xi})$ is computed using the analysis relationship:

$$\hat{y}_k = \int_{\Omega} y(\boldsymbol{\xi}) H_k(\boldsymbol{\xi}) p(\boldsymbol{\xi}) d\boldsymbol{\xi},$$

which can be approximated as

$$\hat{y}_k = \sum_{j=1}^{\hat{N}} y(\boldsymbol{\xi}^j) H_k(\boldsymbol{\xi}^j) w^j \tag{7.1}$$

where the sampling space of the parameters and their weights in Eq. (7.1), $\left\{\boldsymbol{\xi}^j, w^j\right\}$, are selected from a quadrature rule. $y(\boldsymbol{\xi}^j)$ is determined by solving the deterministic problem $\mathcal{F}((\mathbf{x}(t, \boldsymbol{\xi}^j), \boldsymbol{\xi}^j) = 0$ for each realization of the random variables.

7.3.2.1 Stochastic Collocation Algorithm

The stochastic collocation simulation method approximates the unknown stochastic solution by a polynomial interpolation function in the parameter ensemble space. The samples of the solution are collected by solving the deterministic problem $\mathcal{L}(\mathbf{x}, t, \xi; u) = f(\mathbf{x}, t; \xi)$ at a predetermined set of points $\Theta_N = \left\{\xi^j\right\}_{j=1}^{N}$. The sampling Θ_N is chosen as a sparse grid using the Smolyak algorithm [13] instead of generating it from its distribution as in the Monte Carlo simulation. The stochastic collocation method is presented in Algorithm 7.2 below [12].

7.4 Monte Carlo-Based UQ Results for Tapered PEH

In this section, we will discuss the UQ results of some target uncertainty measures that help with the design of PEH devices obtained from Monte Carlo simulations. The MC-based UQ shown in Algorithm 7.1 was invoked on the tapered beam PEH device used by Yang and Tang in [14]. For this analysis, the thickness and Young's modulus of the piezoelectric layer, PZT were set up as uncertain parameters to the model. The PZT thickness was allowed to vary according to a Gaussian PDF, p_{h_p}, with the mean equal to its nominal design value of 0.4mm and the standard deviation equal to a third of its process tolerance which is considered to be 10% of

Algorithm 7.2 Stochastic collocation UQ of a tapered PEH

procedure SCUQOFTPEH(*ModalAnalysis, SteadyStateAnalysis,* and *BuildECM*) *ModalAnalysis*: Algorithm 4.1, *SteadyStateAnalysis*: Algorithm 6.1

Prepare ensemble space

Identify independent design parameters with variability impacting device performance:

$$\boldsymbol{\xi} = \left[\xi^1, \xi^2, \ldots, \xi^n\right]^T \text{ and } \boldsymbol{\xi} \in \Omega.$$

Generate the sparse grid nodes $\Theta_N = \left\{\boldsymbol{\xi}^j\right\}_{j=1}^N$ using Smolyak algorithm [13] based on an appropriate 1-D interpolation rule.

Invoke device models

Invoke *ModalAnalysis*$\forall \boldsymbol{\xi}^j$ to compute f_r, and $\bar{\kappa}_r$

Invoke *SteadyStateAnalysis* $\forall \boldsymbol{\xi}^j$ to compute $P_{r,max}$, $V_{r,max}$, and $w_{\max}$

Invoke *BuildECM*$\forall \boldsymbol{\xi}^j$ to compute C_p, R_r, C_r, N_r, fg_r

Construct interpolant for PDFs

For each of the sampled quantities computed above $u(x, t, \boldsymbol{\xi}^j)$, construct the interpolant as

$$\hat{u}(x, t, \boldsymbol{\xi}) = \sum_{j=1}^{N} u(x, t, \boldsymbol{\xi}^j) L_j(\boldsymbol{\xi}),$$

where $\left\{L_j(\boldsymbol{\xi})\right\}_{j=1}^N$ are the interpolation basis functions.

Compute Moments

Compute the required statistics as

$$\mu(u)(x, t) = \sum_{j=1}^{N} u(x, t, \boldsymbol{\xi}^j) w_j, \text{ and}$$

$$\sigma^2(u)(x, t) = \sum_{j=1}^{N} \left(u(x, t, \boldsymbol{\xi}^j) - \mu(u)\right)^2 w_j.$$

where $w_j = \int_\Omega L_j(\boldsymbol{\xi}) p(\boldsymbol{\xi}) d\boldsymbol{\xi}$

Compute PDFs

Generate i.i.d. random variables $\left\{\boldsymbol{\eta}^j\right\}_{j=1}^{N_{mc}}$ for $N_{mc} \gg N$.

For each quantity sampled above, evaluate interpolant at $\boldsymbol{\eta}$ as $\hat{u}^j = \hat{u}(x, t, \boldsymbol{\eta})$.

Determine the pdf from the interpolant set $\left\{\hat{u}^j\right\}_{j=1}^N$.

end procedure

its design value. PZT Young's modulus was also considered to be a random variable with a Gaussian PDF of p_{Y_p} with its mean at the declared modulus value and the standard deviation a third of the 10% tolerance. The standard deviation of the input uncertainty is, therefore, set up at about 3.3% of the nominal design values.

7.4.1 *Modal Analysis Uncertainty Measures*

Modal analysis results directly impact design flow in two important ways. The resonance frequency values obtained from the modal analysis are crucial in ascertaining the target operational frequency of the device. Resonance frequency estimates also have significant downstream effects on the device model. A second output of the modal analysis is the mode shapes. Mode shapes affect the computation of bending moments of the structure for the governing system through their influence on the coupling factor, the actuation term, and the damping coefficients. Moreover, the stress distribution values obtained from the mode-shape functions can help gauge the structural integrity of the device as well as the energy harvester output efficiency.

7.4.1.1 Resonance Frequency

Figure 7.1 shows the histogram estimate of the PDFs of the first mode resonance frequency p_{f_1} and Fig. 7.2 shows the PDF estimate of the second mode resonance frequency, p_{f_2}. The PDFs of the resonance frequencies remain near Gaussian. The standard deviation of the higher-frequency second mode is higher than that of the first mode. The standard deviation value for the first mode resonance frequency of 1.09 is about 1.5% of the nominal value of 65 Hz, which means that the variability introduced in the PZT thickness and its Young's modulus has scaled down as it reaches the resonance frequency output. With a standard deviation of only 1.09 Hz, the yield in regards to meeting the frequency target may be high, but we know

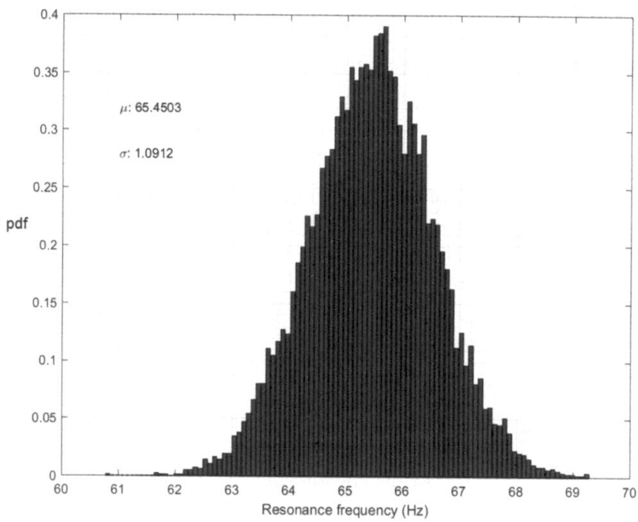

Fig. 7.1 Histogram estimate of the PDF of the first mode resonance frequency

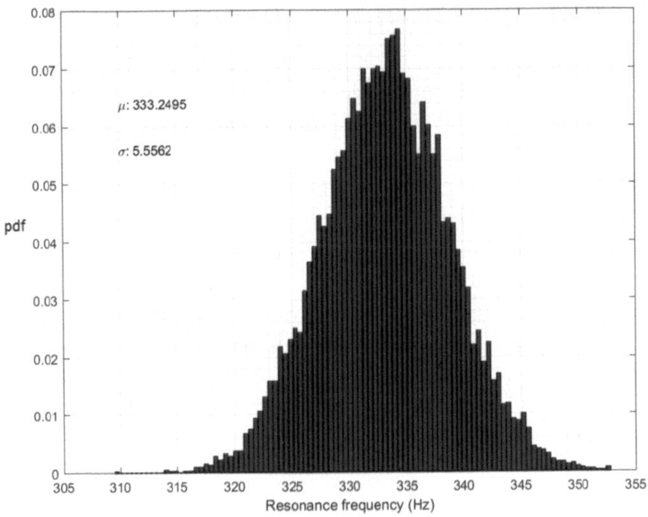

Fig. 7.2 Histogram estimate of the PDF of the second mode resonance frequency

that a slight resonance frequency variability will affect the maximum output power variability significantly.

7.4.1.2 Average Curvature of the Mode Shape

We evaluated the normalized average curvature $\bar{\kappa}$ of the mode shapes as a measure of the structural health of the device. A $\bar{\kappa}$ value of unity implies the curvature is perfectly spread along the beam such that the average curvature is the same as the maximum curvature. A small value implies the curvature is concentrated at specific points on the beam. Figure 7.3 shows the PDF distribution of the normalized average curvature of Yang's tapered beam example. The small standard deviation of 4×10^{-12} means that the variability in the piezoelectric layer thickness and its Young's modulus does not translate into much variability in the curvature of the beam.

7.4.2 Steady-State Analysis Uncertainty Measures

Steady-state (AC) analysis outputs that influence the design of energy harvesters are the maximum output power along with the output voltage and the maximum beam tip deflection. Variability study of maximum output power is of critical importance. Any contributing factors to a reduction in expected power output must be mitigated. Maximum beam deflection helps determine the saturation points of capped devices

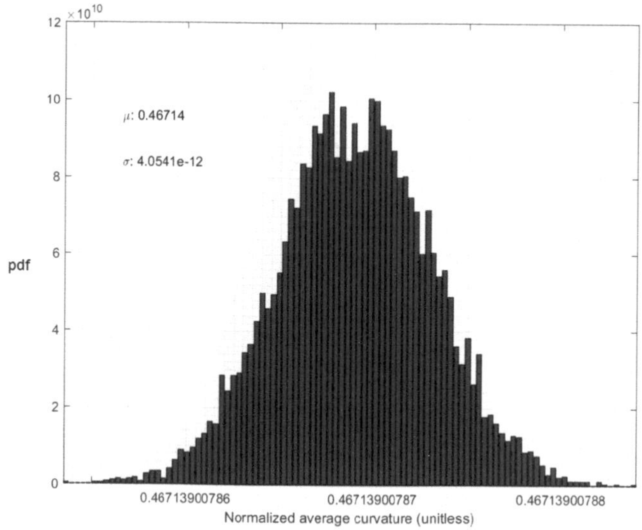

Fig. 7.3 PDF of normalized average curvature $\bar{\kappa}$ for the first mode

where higher beam deflection results in touching the caps above and below the device.

7.4.2.1 Maximum Output Power Results at Optimal Load

We computed the output power for a range of load situations and computed the maximum power at the optimal load for each of the modes. The histogram estimate of the PDF of the optimal load computed using MC is shown in Fig. 7.4. A standard deviation of about 0.75 kΩ is about 3.4% of the nominal R_{opt} value. We know that the maximum power is measured at the optimal load. Whereas a higher uncertainty in R_{opt} adds to characterization challenges, our UQ results for maximum output power are not affected by R_{opt} variability, since each $P_{\max}$ is computed at its respective R_{opt}.

The histogram estimate of the PDF of $P_{\max}$ is shown in Fig. 7.5. The standard deviation is about 1% of the nominal value. This is low because $P_{\max}$ is evaluated at the resonance frequency for the optimal load and in doing so its PDF has been conditioned on the PDFs of resonance frequency and the optimal load. We expect that the output power at a specified frequency for a specified load would have a greater variability.

The maximum output voltages are also computed at the same optimal load used for maximum power. The UQ results computed using MC simulations are shown in Fig. 7.6.

The maximum beam deflection for the first mode was computed using Eq. (6.25). The UQ result is shown in Fig. 7.7.

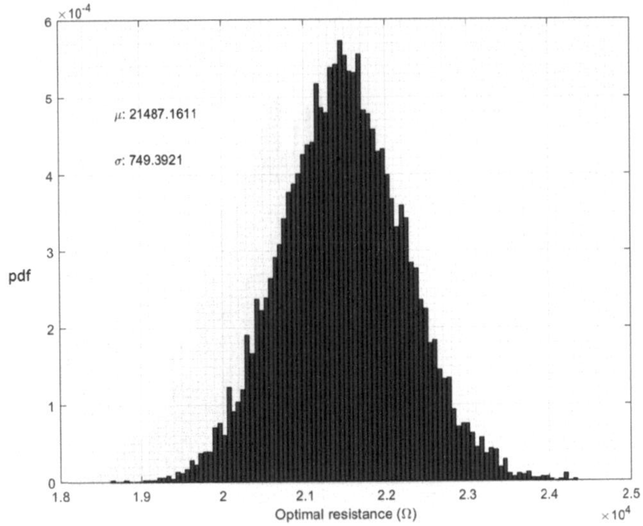

Fig. 7.4 Histogram estimate of the PDF of the optimum load

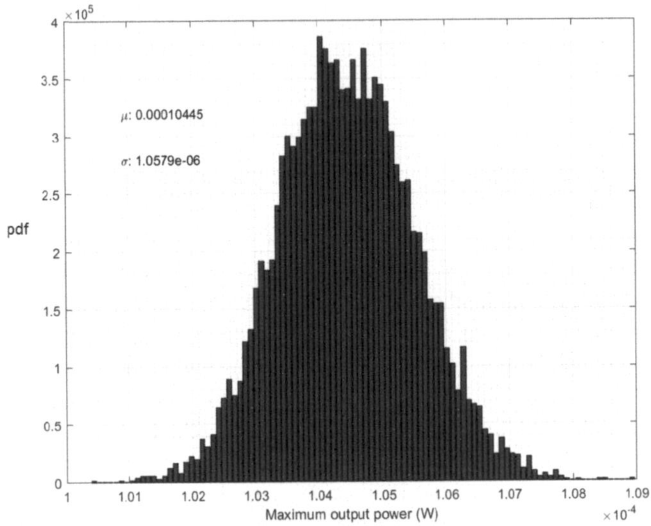

Fig. 7.5 Histogram estimate of the PDF of $P_{\max}$ for the first mode

7.5 ECM and ECM Response UQ Results

The ECM of the tapered beam energy harvester was established in Sect. 6.4. In this section, we will discuss the variability of each of the ECM components when the PZT thickness is uncertain. As before, we use Yang's simple tapered beam EH to

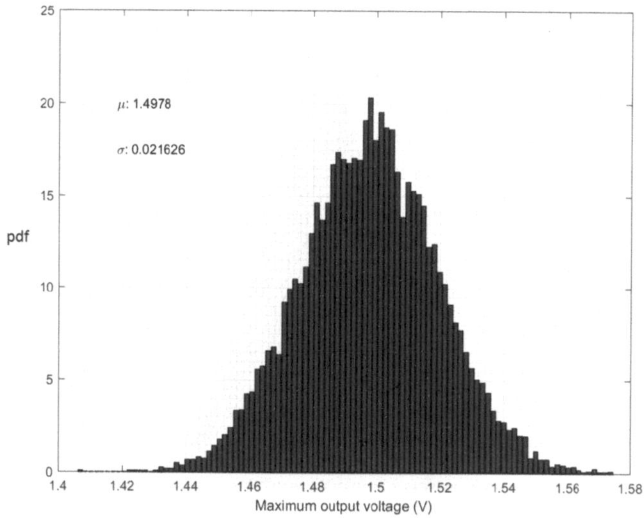

Fig. 7.6 Histogram estimate of the PDF of V_{max} for the first mode

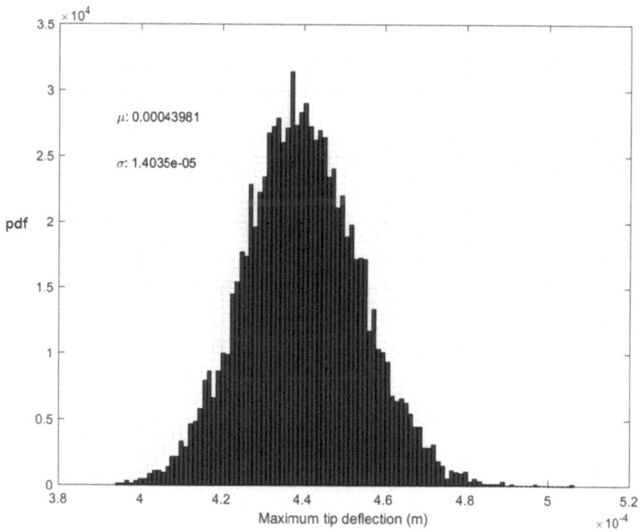

Fig. 7.7 Histogram estimate of the PDF of w_{max} of the first mode

illustrate the methodology. The piezoelectric capacitance is inversely proportional to the piezoelectric layer thickness. The UQ of C_p is shown in Fig. 7.8 suggesting that the standard deviation in C_p is about 3.3% of its nominal value which is about the same as the PZT thickness standard deviation.

The histogram estimates of the PDFs of R_1, C_1, N_1, and fg_1 are shown in Figs. 7.9, 7.10, 7.11, 7.12. From among the ECM parameters, we know that C_1 is

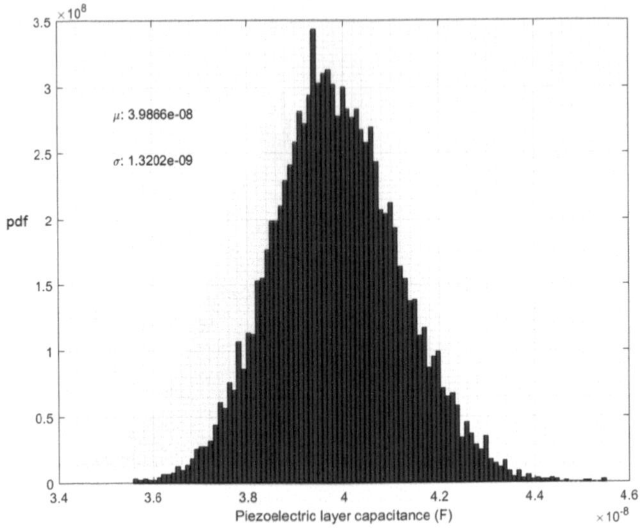

Fig. 7.8 PDF of C_p computed using MC simulations for uncertain h_p

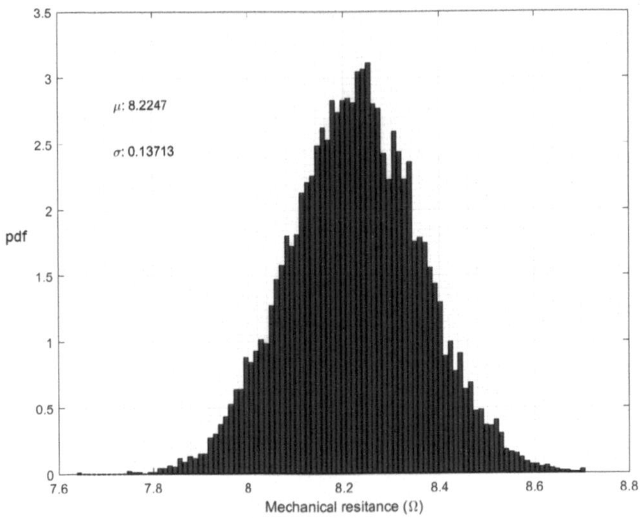

Fig. 7.9 Histogram estimate of the PDF of R_1

the most affected by changes in PZT layer thickness and Young's modulus because of their effect on the compliance of the beam. We notice from the results that C_1 has a standard deviation of more than 3.2% of its nominal value, whereas the standard deviation percentage value for the other parameters is lower than 1.5%.

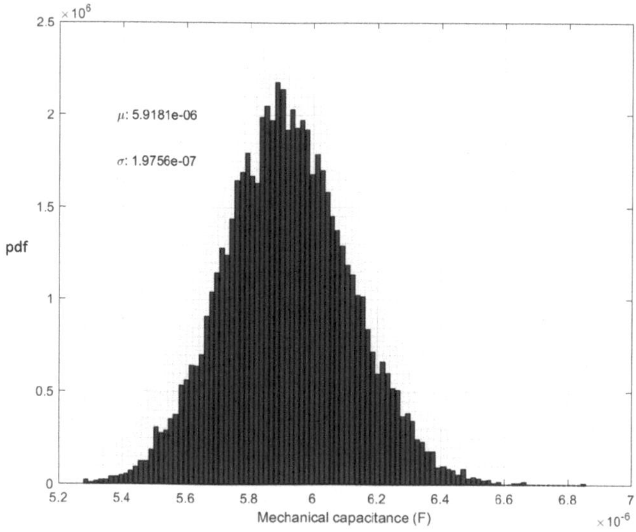

Fig. 7.10 Histogram estimate of the PDF of C_1

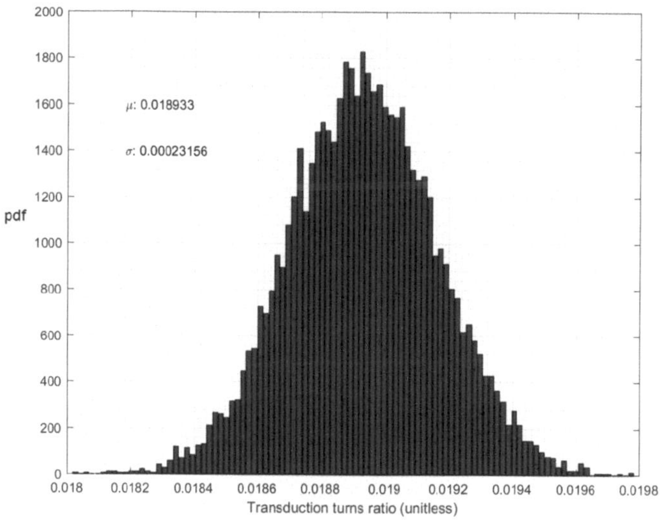

Fig. 7.11 Histogram estimate of the PDF of N_1

7.6 Conclusion

In this chapter, we have presented the uncertainty quantification results of various parameters and outputs of the tapered PEH model using Monte Carlo simulation. We demonstrated that the perturbation expansion-based tapered PEH model developed

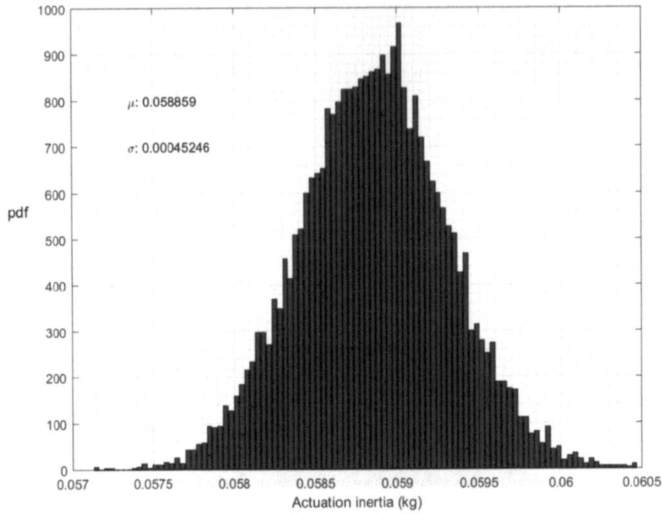

Fig. 7.12 Histogram estimate of the PDF of fg_1

in this work is well suited for Monte Carlo analysis owing to its high speed of simulation when compared to traditional FEM-based models.

References

1. Dongbin Xiu and George Em Karniadakis. "Modeling uncertainty in flow simulations via generalized polynomial chaos". In: *Journal of computational physics* 187.1 (2003), pp. 137–167.
2. Nitin Agarwal and NR Aluru. "A data-driven stochastic collocation approach for uncertainty quantification in MEMS". In: *International Journal for Numerical Methods in Engineering* 83.5 (2010), pp. 575–597.
3. Aravind Alwan and NR Aluru. "Uncertainty quantification of MEMS using a data-dependent adaptive stochastic collocation method". In: *Computer Methods in Applied Mechanics and Engineering* 200.45 (2011), pp. 3169–3182.
4. James W Tschanz et al. "Adaptive body bias for reducing impacts of die-to-die and within-die parameter variations on microprocessor frequency and leakage". In: *IEEE Journal of Solid-State Circuits* 37.11 (2002), pp. 1396–1402.
5. Jonathan W Wittwer, Troy Gomm, and Larry L Howell. "Surface microma-chined force gauges: uncertainty and reliability". In: *Journal of Micromechanics and Microengineering* 12.1 (2001), p. 13.
6. Stefan Reh, Paul Lethbridge, and Dale Ostergaard. "Quality based design and design for reliability of micro electro mechanical systems (MEMS) using probabilistic methods". In: *2000 International Conference on Modeling and Simulation of Microsystems, San Diego, CA*. 2000.
7. JS Kong et al. "A methodology for analyzing the variability in the performance of a MEMS actuator made from a novel ceramic". In: *Sensors and Actuators A: Physical* 116.2 (2004), pp. 336–344.

8. Matthew Allen et al. “Reliability-based analysis and design optimization of electrostatically actuated MEMS”. In: *Computers & structures* 82.13 (2004), pp. 1007–1020.
9. Jeong Sam Han and Byung Man Kwak. “Robust optimal design of a vibratory microgyroscope considering fabrication errors”. In: *Journal of micromechanics and microengineering* 11.6 (2001), p. 662.
10. Rong Liu, Brad Paden, and Kimberly Turner. “MEMS resonators that are robust to process-induced feature width variations”. In: *Journal of Microelectromechanical Systems* 11.5 (2002), pp. 505–511.
11. Zheng Zhang. “Uncertainty Quantification for Integrated Circuits and Microelectromechanical Systems”. PhD thesis. Citeseer, 2015.
12. Nitin Agarwal and Narayana R Aluru. “Stochastic analysis of electrostatic MEMS subjected to parameter variations”. In: *Journal of Microelectromechanical Systems* 18.6 (2009), pp. 1454–1468.
13. Sergey A Smolyak. “Quadrature and interpolation formulas for tensor products of certain classes of functions”. In: *Dokl. Akad. Nauk SSSR*. Vol. 4. 1963, pp. 240–243.
14. Yaowen Yang and Lihua Tang. “Equivalent circuit modeling of piezoelectric energy harvesters”. In: *Journal of Intelligent Material Systems and Structures* (2009), pp. 2223–2235.

Chapter 8
Uncertainty Quantification Using Polynomial-Chaos Expansion

In this chapter, we explore uncertainty quantification techniques based on polynomial chaos expansion. We begin by looking at the general MEMS stochastic equation before setting up the stochastic governing system for the tapered piezoelectric energy harvester. We first illustrate the application of polynomial chaos expansion on the general MEMS stochastic equation. We then proceed to apply the PCE on the tapered PEH stochastic equation and apply Galerkin projection on the expanded system for its solution. We present the uncertainty dependency graph of the tapered PEH model that can be used to incrementally propagate the uncertainty model through to the desired node. We use a numerical probability transformation approach to estimate the probability density functions at any of the graph nodes. In the end, we propose to use the stochastic testing-based hierarchical UQ approach developed by Zhang [1] on the TPEH equivalent circuit model.

8.1 The Stochastic Governing System

To set up the problem for polynomial chaos expansion-based uncertainty quantification of MEMS devices, the governing system of the device needs to be formulated along with the uncertainties, often leading to stochastic differential equations. A general MEMS stochastic equation is presented in Sect. 8.1.1. The stochastic governing system of the tapered beam piezoelectric energy harvester is given in Sect. 8.1.2. The tapered beam EH system is mapped onto an equivalent circuit model which can be used instead for uncertainty quantification using the UQ methodology worked out for circuits by Zhang [2].

W. U. Syed, I. (Abe) M. Elfadel, *Tapered Beams in MEMS*,
https://doi.org/10.1007/978-3-031-66391-8_8

8.1.1 General MEMS Stochastic Equation

A lumped model of a MEMS device sets up a mass–spring–dashpot system and can be represented along with uncertainties in the lumped parameters by the following second-order stochastic equation:

$$M(\mathbf{z}, \boldsymbol{\xi}) \frac{d^2\mathbf{z}}{dt^2} + D(\mathbf{z}, \boldsymbol{\xi}) \frac{d\mathbf{z}}{dt} + \tilde{f}(\mathbf{z}, \mathbf{u}, \boldsymbol{\xi}) = 0, \tag{8.1}$$

where $\boldsymbol{\xi} \in \Omega \subseteq \mathbb{R}^d$ is the d-dimensional random variable that represents process variation; $\mathbf{z}(t, \boldsymbol{\xi}) \in \mathbb{R}^n$, denoting the displacements and rotations, is a random process in that it is a function of t and $\boldsymbol{\xi}$; $\mathbf{u}$ is a function of t and represents the inputs to the system such as voltage sources, mechanical forces, or acceleration; $M, D \in R^{n \times n}$ are the mass matrix and damping coefficient matrix, respectively; and $\tilde{f}$ denotes the net forces from all actuation sources. For simplicity, we will assume independence between all random parameters and inputs which means that their joint probability density function can be expressed as

$$p(\boldsymbol{\xi}) = \prod_{k=1}^{d} p_k(\xi_k),$$

where $p_k(\xi_k)$ is the probability density function of $\xi_k \in \Omega_k$. The structural dynamics in Eq. (8.1) couples with a range of other governing equations depending on the active physics in the device. In the case of an energy harvester, it couples with an electric circuit equation developed in Sect. 6.1.3 that is derived from the piezoelectric constitutive system.

A typical state-space system constitutes a state update equation and an output equation. Let us assume that the output of interest $y(t)$ is computed from the uncertain states $\mathbf{x}(t, \boldsymbol{\xi}) \in \mathbb{R}^n$ with an operator $\mathcal{Y}$ and the states satisfy the system governing set of equations defined by $\mathcal{F}$:

$$\mathcal{F}(\mathbf{x}, \boldsymbol{\xi}) = 0, \text{ and}$$
$$y = \mathcal{Y}(\mathbf{x}).$$

The 2nd-order stochastic differential equation, Eq. (8.1), can be converted into a first-order form for state-space representation by choosing an appropriate vector for the system states:

$$\mathbf{x} = \begin{pmatrix} \mathbf{z} \\ \frac{d\mathbf{z}}{dt} \end{pmatrix},$$

and defining the forcing function as

$$\mathbf{f} = \begin{pmatrix} \tilde{f} \\ 0 \end{pmatrix}$$

and simplifying the state derivative term as

$$\frac{d\mathbf{q}}{dt} = \begin{pmatrix} D & M \\ 0 & I \end{pmatrix} \frac{d\mathbf{x}}{dt}$$

to get a stochastic differential-algebraic equation (DAE) for the system equation as

$$\frac{d}{dt}\mathbf{q}(\mathbf{x}, \boldsymbol{\xi}) + \mathbf{f}(\mathbf{x}, \mathbf{u}, \boldsymbol{\xi}) = 0,$$

which can be solved using stochastic circuit solvers.

8.1.2 *Stochastic Tapered Beam Piezoelectric EH Governing System*

The deterministic electromechanical governing system of a tapered TPEH derived in Sect. 6.3.2 is a differential equation system of the modal function $\eta_r(t)$ and the instantaneous voltage $v(t)$ given as

$$\ddot{\eta}_r(t) + 2\zeta_r\omega_r\dot{\eta}_r(t) + \eta_r(t)\omega_r^2 + \chi_r v(t) = -F_r(t), \text{ and}$$

$$-\sum_{r=1}^{\infty} \chi_r\dot{\eta}_r(t) + \frac{1}{R_l}v(t) + C_p\dot{v}(t) = 0.$$

The lumped parameters in these equations, ω_r, χ_r, and C_p can be defined in terms of the uncertainties that they carry as ω_r^{ξ}, χ_r^{ξ}, and C_p^{ξ} to get

$$\ddot{\eta}_r(t, \boldsymbol{\xi}) + 2\zeta_r^{\xi}\omega_r^{\xi}\dot{\eta}_r(t, \boldsymbol{\xi}) + \eta_r(t, \boldsymbol{\xi})(\omega_r^{\xi})^2 + \chi_r^{\xi} v(t, \boldsymbol{\xi}) = -F_r(t, \boldsymbol{\xi}), \text{ and}$$

$$-\sum_{r=1}^{\infty} \chi_r^{\xi}\dot{\eta}_r(t, \boldsymbol{\xi}) + \frac{1}{R_l}v(t, \boldsymbol{\xi}) + C_r^{\xi}\dot{v}(t, \boldsymbol{\xi}) = 0.$$

This can be expressed as a stochastic state-space system setting up the states as in Sect. 6.3.2 as

$$\frac{d\mathbf{q}(t, \boldsymbol{\xi})}{dt} = \boldsymbol{A}(\boldsymbol{\xi})\mathbf{q}(t, \boldsymbol{\xi}) + \boldsymbol{B}(\boldsymbol{\xi})\mathbf{u}(t), \tag{8.2}$$

where the states are arranged as

$$\mathbf{q}(t,\boldsymbol{\xi}) = \begin{bmatrix} \dot{\eta}_1(t,\boldsymbol{\xi}) \\ \eta_1(t,\boldsymbol{\xi}) \\ \dot{\eta}_2(t,\boldsymbol{\xi}) \\ \eta_2(t,\boldsymbol{\xi}) \\ \vdots \\ \dot{\eta}_{N_m}(t,\boldsymbol{\xi}) \\ \eta_{N_m}(t,\boldsymbol{\xi}) \\ v(t,\boldsymbol{\xi}) \end{bmatrix},$$

and the acceleration input vector is

$$\mathbf{u}(t) = \begin{bmatrix} \ddot{g}(t) \\ \ddot{h}(t) \end{bmatrix}.$$

The matrices $\boldsymbol{A}(\boldsymbol{\xi})$ and $\boldsymbol{B}(\boldsymbol{\xi})$ are given as

$$\mathbf{A}(\boldsymbol{\xi}) = \begin{bmatrix} -2\zeta_1^{\xi}\omega_1^{\xi} & -(\omega_1^{\xi})^2 & 0 & 0 & \cdots & 0 & 0 & -\chi_1^{\xi} \\ 1 & 0 & 0 & 0 & \cdots & 0 & 0 & 0 \\ 0 & 0 & -2\zeta_2^{\xi}\omega_2^{\xi} & -(\omega_2^{\xi})^2 & \cdots & 0 & 0 & -\chi_2^{\xi} \\ 0 & 0 & 1 & 0 & \cdots & 0 & 0 & 0 \\ \vdots & \vdots & \vdots & \vdots & \vdots & \vdots & \vdots & \vdots \\ 0 & 0 & 0 & 0 & \cdots & 2\zeta_{N_m}^{\xi}\omega_{N_m}^{\xi} & -(\omega_{N_m}^{\xi})^2 & -\chi_{N_m}^{\xi} \\ 0 & 0 & 0 & 0 & \cdots & 1 & 0 & 0 \\ \frac{\chi_1^{\xi}}{C_p^{\xi}} & 0 & \frac{\chi_2^{\xi}}{C_p^{\xi}} & 0 & \cdots & \frac{\chi_{N_m}^{\xi}}{C_p^{\xi}} & 0 & -\frac{1}{R_l C_p^{\xi}} \end{bmatrix},$$

and

$$\mathbf{B}(\boldsymbol{\xi}) = \begin{bmatrix} -P_0^m(\phi_1(\boldsymbol{\xi})) & -P_1^m(\phi_1(\boldsymbol{\xi})) \\ 0 & 0 \\ -P_0^m(\phi_2(\boldsymbol{\xi})) & -P_1^m(\phi_2(\boldsymbol{\xi})) \\ 0 & 0 \\ \vdots & \vdots \\ -P_0^m(\phi_1(\boldsymbol{\xi})) & -P_1^m(\phi_1(\boldsymbol{\xi})) \\ 0 & 0 \\ 0 & 0 \end{bmatrix}.$$

Assuming that the modes are well separated, we simplify this system to solve for one mode at a time. The system vectors and matrices for a single mode are simplified to

$$\mathbf{q}(t,\boldsymbol{\xi}) = \begin{bmatrix} \dot{\eta}(t,\boldsymbol{\xi}) \\ \eta(t,\boldsymbol{\xi}) \\ v(t,\boldsymbol{\xi}) \end{bmatrix},$$

$$\mathbf{A}_r(\boldsymbol{\xi}) = \begin{bmatrix} -2\zeta_r^{\xi}\omega_r^{\xi} & -(\omega_r^{\xi})^2 & \chi_r^{\xi} \\ 1 & 0 & 0 \\ \frac{\chi_r^{\xi}}{C_p^{\xi}} & 0 & -\frac{1}{R_l C_p^{\xi}} \end{bmatrix}, \text{ and}$$

$$\mathbf{B}_r(\boldsymbol{\xi}) = \begin{bmatrix} -P_0^m(\phi_r(\boldsymbol{\xi})) & -P_1^m(\phi_r(\boldsymbol{\xi})) \\ 0 & 0 \\ 0 & 0 \end{bmatrix}.$$

8.2 Polynomial Chaos Expansion of the MEMS Stochastic Equation

Spectral methods for uncertainty quantification aim at describing the solution y on the random input (or model parameter) ξ as

$$y(\xi) = \sum_{i=0}^{\infty} y_i \phi_i(\xi),$$

where ϕ_i for $i = 0, \ldots, \infty$ are selected basis functions and y_i denote the deterministic *modes* of y. Polynomial chaos expansion uses polynomials for the basis functions in spectral expansion. Wiener first introduced PCE [3] using Hermite-type polynomials to approximate $\mathbf{x}(t,\boldsymbol{\xi})$ with normally distributed parameters. Generalized polynomial chaos, introduced by Xiu and Karniadiakis in 2002 [4], expands on Wiener's work and helps extend the notion to non-Gaussian random parameters. For smooth $\mathbf{x}(t,\boldsymbol{\xi})$ with a bounded 2nd-order moment, $\mathbf{x}(t,\boldsymbol{\xi})$ can be approximated by a finite-term generalized polynomial expansion as

$$\mathbf{x}(t,\boldsymbol{\xi}) \approx \tilde{x}(t,\boldsymbol{\xi}) = \sum_{\boldsymbol{\alpha}\in\mathcal{P}} \hat{x}_{\boldsymbol{\alpha}}(t) H_{\boldsymbol{\alpha}}(\boldsymbol{\xi}),$$

where $H_{\boldsymbol{\alpha}}(\boldsymbol{\xi})$ is a basis function indexed by $\boldsymbol{\alpha}$, $\hat{x}_{\boldsymbol{\alpha}}(t) \in \mathbb{R}^n$ used to represent the modes of $\mathbf{x}(t,\boldsymbol{\xi})$, and $\mathcal{P}$ is a well-designed index set. For simplicity when approximating using a finite number P of polynomials, this is also written as

$$\mathbf{x}(t, \boldsymbol{\xi}) \approx \tilde{x}(t, \boldsymbol{\xi}) = \sum_{k=1}^{P} \hat{x}_k(t) H_k(\boldsymbol{\xi}).$$

We choose the value of the number of terms P to keep in the approximation using the formula [5]:

$$(P + 1) = \frac{(n + m)!}{(n!m!)},$$

where n is the number of random variables and m is the maximum degree of the polynomials.

8.2.1 Finding Orthogonal Polynomials

The basis functions for gPC are chosen such that the orthonormality condition,

$$\left\langle \phi_\nu^k(\xi_k), \phi_{\nu'}^k(\xi_k) \right\rangle_{\Omega_k, p_k(\xi_k)} = \int_{\Omega_k} p_k(\xi_k) \phi_{\nu'}^k(\xi_k) \phi_\nu^k(\xi_k) d\xi_k = \delta_{\nu,\nu'},$$

holds, where $\langle \cdot, \cdot \rangle_{\Omega_k, p_k(\xi_k)}$ is the inner product with respect to the density function $p_k(\xi_k)$ over the space Ω_k, $\delta_{\nu,\nu'}$ is the Dirac delta function, and the indices ν and ν' are the highest degree of ξ_k in the polynomials.

Gautschi suggested the following procedure [6] for determining the polynomials $\left\{\phi_\nu^k(\xi_k)\right\}_{\nu=0}^{m}$ much like the Gram–Schmidt procedure for obtaining orthogonal bases in a vector space.

1. Use the following recurrence relationship to derive a set of polynomials $\left\{\pi_\nu^k(\xi_k)\right\}_{\nu=0}^{m}$ with a unity leading coefficient

$$\pi_{\nu+1}^k(\xi_k) = (\xi_k - \gamma_\nu)\pi_\nu^k(\xi_k) - \kappa_\nu \pi_{\nu-1}^k(\xi_k), \ \nu = 0, 1, \cdots, m - 1,$$

with initial conditions $\pi_{-1}^k(\xi_k) = 0$, $\pi_0^k(\xi_k) = 1$,$\kappa_0 = 1$, and for $\nu \geq 0$

$$\gamma_\nu^k = \frac{\mathbb{E}\left(\xi_k(\pi_\nu^k(\xi_k))^2\right)}{\mathbb{E}\left((\pi_\nu^k(\xi_k))^2\right)}, \text{ and}$$

$$\kappa_{\nu+1}^k = \frac{\mathbb{E}\left(\xi_k(\pi_{\nu+1}^k(\xi_k))^2\right)}{\mathbb{E}\left(\xi_k(\pi_\nu^k(\xi_k))^2\right)},$$

where $\mathbb{E}$ denotes the expectation operator.

Table 8.1 Correspondence of probability distribution of the random variables with the Wiener–Askey chaos polynomial [5]

	Probability distribution of ξ	Wiener–Askey chaos polynomial $H(\xi)$	Support
Continuous	Gaussian	Hermite-chaos	$(-\infty, \infty)$
	Gamma	Laguerre-chaos	$[0, \infty)$
	Beta	Jacobi-chaos	$[a, b]$
	Uniform	Legendre-chaos	$[a, b]$
Discrete	Poisson	Charlier-chaos	$\{0, 1, 2, \cdots\}$
	Binomial	Krawtchouk-chaos	$\{0, 1, 2, \cdots, N\}$
	Negative binomial	Meixner-chaos	$\{0, 1, 2, \cdots\}$
	Hypergeometric	Hahn-chaos	$\{0, 1, 2, \cdots, N\}$

2. Find the orthogonal polynomial $\left\{\phi_\nu^k(\xi_k)\right\}_{\nu=0}^m$ after normalizing the polynomials $\pi_\nu^k(\xi)$ to get

$$\phi_\nu^k(\xi_k) = \frac{\pi_\nu^k(\xi_k)}{\sqrt{\kappa_0 \kappa_1 \cdots \kappa_\nu}}, \quad \nu = 0, 1, \cdots, m.$$

Bases for multivariate generalized polynomial chaos when ξ_k are assumed independent are derived from the univariate procedure given above as

$$H_{\boldsymbol{\alpha}}(\boldsymbol{\xi}) = \prod_{k=1}^{d} \phi_{\alpha_k}^k(\xi_k),$$

where $\boldsymbol{\alpha} = [\alpha_1, \cdots, \alpha_d] \in \mathbb{N}^d$ is a multidimensional index vector.

Based on the probability distribution of the independent random variables, we can use Table 8.1 to determine the corresponding chaos polynomial.

8.2.1.1 Calculating Expansion Modes

The coefficient functions $\hat{x}_k(t)$, also called the modes of expansion, are given by the inner product of $\mathbf{x}(t, \boldsymbol{\xi})$ with the polynomial kth. Given that the index vector $\boldsymbol{\alpha}$ maps onto an index k where $0 \leq k \leq P$, the coefficient functions can be written in terms of the polynomial chaos functions as

$$\hat{x}_k(t) = \langle \mathbf{x}(t, \boldsymbol{\xi}), H_k(\boldsymbol{\xi}) \rangle .$$

8.2.1.2 Statistics of the Solution

Generalized polynomial chaos makes it very convenient to find statistics of the solution $\mathbf{x}(t, \boldsymbol{\xi})$. We note that the expected values of the polynomials are conditioned as

$$\mathbb{E}\left(H_0(\boldsymbol{\xi})\right) = 1, \text{ and} \tag{8.3}$$

$$\mathbb{E}\left(H_k(\boldsymbol{\xi})\right)|_{k>0} = 0. \tag{8.4}$$

Using Eqs. (8.3) and (8.4), the mean value of the $\mathbf{x}(t, \boldsymbol{\xi})$ is calculated as

$$\mathbb{E}\left(\mathbf{x}(t, \boldsymbol{\xi})\right) = \hat{x}_{\alpha=0}(t), \tag{8.5}$$

and the standard deviation of $\mathbf{x}(t, \boldsymbol{\xi})$ is calculated as

$$\sigma\left(\mathbf{x}(t, \boldsymbol{\xi})\right) = \sqrt{\sum_{\alpha \neq 0} \left|\hat{x}_\alpha(t)\right|^2}.$$

8.3 Galerkin Projection on PCE

Stochastic Galerkin is an intrusive solver that uses polynomial chaos functions, $\{H_k(\boldsymbol{\xi})\}_{k=0}^{P}$, as its basis functions and applies a Galerkin projection to set up a deterministic model that is solved one time to evaluate the states and consequently the outputs. Stochastic Galerkin is, in effect, the generalization of the finite-element method to the parameter sample space.

8.3.1 General Procedure to Galerkin Projection PCE

Let us consider the stochastic differential equation:

$$\mathcal{L}\left(\mathbf{x}, t, \boldsymbol{\xi}; u\right) = f\left(\mathbf{x}, t; \xi\right), \tag{8.6}$$

where $u(\mathbf{x}, t; \boldsymbol{\xi})$ is the solution and $f(\mathbf{x}, t; \boldsymbol{\xi})$ is the forcing function on the system. The operator $\mathcal{L}$ describes the governing differential equation of the system. Let us expand the solution u by PCE as

$$u(\mathbf{x}, t; \boldsymbol{\xi}) = \sum_{k=0}^{P} u_k(\mathbf{x}, t) H_k(\boldsymbol{\xi}). \tag{8.7}$$

We substitute $u(\mathbf{x}, t; \boldsymbol{\xi})$ from Eq. (8.7) to Eq. (8.6) to get [5]

$$\mathcal{L}\left(\mathbf{x}, t, \boldsymbol{\xi}; \sum_{k=0}^{P} u_k(\mathbf{x}, t) H_k(\boldsymbol{\xi})\right) = f(\mathbf{x}, t; \boldsymbol{\xi}). \tag{8.8}$$

We then perform Galerkin projection of Eq. (8.8) onto each polynomial basis H_i:

$$\left\langle \mathcal{L}\left(\mathbf{x}, t, \boldsymbol{\xi}; \sum_{k=0}^{P} u_k(\mathbf{x}, t) H_k(\boldsymbol{\xi})\right), H_i(\boldsymbol{\xi})\right\rangle = \langle f(\mathbf{x}, t; \boldsymbol{\xi}), H_i(\boldsymbol{\xi})\rangle, \; i = 0, 1, \cdots, P. \tag{8.9}$$

We use orthogonality between the polynomial bases to reduce Eq. (8.9) to a set of $P + 1$ deterministic simultaneous equations for each random mode $u_k(\mathbf{x}, t)$ for $k = \{0, 1, \cdots, P\}$.

8.3.2 PCE Applied to Stochastic MEMS Equation

We turn our attention back to the stochastic MEMS equation (Eq. 8.1) to illustrate the working of Galerkin projection. For simplicity, we begin by assuming that the M, D, and K matrices are deterministic. Equation (8.1) then simplifies to

$$M(\xi)\frac{d^2 w(t, \xi)}{dt^2} + D(\xi)\frac{dw(t, \xi)}{dt} + K(\xi)w(t, \xi) = f(t). \tag{8.10}$$

Substituting $w(t; \xi) = \sum_{k=0}^{P} w_k(t) H_k(\xi)$ in Eq. (8.10) we get

$$\left(\sum_{k=0}^{P} M_k H_k(\xi)\right) \frac{d^2}{dt^2} \left(\sum_{k=0}^{P} w_k(t) H_k(\xi)\right)$$
$$\left(\sum_{k=0}^{P} D_k H_k(\xi)\right) \frac{d}{dt} \left(\sum_{k=0}^{P} w_k(t) H_k(\xi)\right)$$
$$\left(\sum_{k=0}^{P} K_k H_k(\xi)\right) \left(\sum_{k=0}^{P} w_k(t) H_k(\xi)\right) = f(t).$$

Taking inner product of both sides with $H_i(\xi)$ and rearranging we get

$$\frac{d^2}{dt^2} \sum_{j=0}^{P} \sum_{k=0}^{P} M_j w_k(t) \langle H_i(\xi) H_j(\xi) H_k(\xi)\rangle$$

Table 8.2 Hermite polynomials for $P = 4$

k	$H_k(\xi)$	$\langle H_k^2(\xi)\rangle$
0	1	1
1	2ξ	1
2	$4\xi^2 - 2$	2
3	$8\xi^3 - 12\xi$	6
4	$16\xi^4 - 48\xi^2 + 12$	24

$$+\frac{d}{dt}\sum_{j=0}^{P}\sum_{k=0}^{P} D_j w_k(t)\big\langle H_i(\xi)H_j(\xi)H_k(\xi)\big\rangle$$

$$+\sum_{j=0}^{P}\sum_{k=0}^{P} K_j w_k(t)\big\langle H_i(\xi)H_j(\xi)H_k(\xi)\big\rangle = f(t)\,\langle H_i(\xi)\rangle\,.$$

Let us define $C_{ijk} = \big\langle H_i(\xi)H_j(\xi)H_k(\xi)\big\rangle$ and substitute to get

$$\sum_{j=0}^{P}\sum_{k=0}^{P} M_j C_{ijk}\frac{d^2}{dt^2}w_k(t) + \sum_{j=0}^{P}\sum_{k=0}^{P} D_j C_{ijk}\frac{d}{dt}w_k(t)$$

$$+\sum_{j=0}^{P}\sum_{k=0}^{P} K_j C_{ijk} w_k(t) = f(t)\,\langle H_i(\xi)\rangle\,. \tag{8.11}$$

Taking into account a single random variable ($n = 1$) that is uniformly distributed and a highest polynomial order of ($m = 4$), we compute $P = \frac{(n+m)!}{(n!m!)} - 1 = 4$ to obtain the set of 5 Hermite polynomials for the truncated approximation shown in Table 8.2.

Expanding Eq. (8.11) and using identities in Eqs. (8.3) and (8.4) leads to the set of deterministic ODEs:

$$\sum_{j=0}^{P}\sum_{k=0}^{P} M_j C_{0jk}\frac{d^2}{dt^2}w_k(t) + \sum_{j=0}^{P}\sum_{k=0}^{P} D_j C_{0jk}\frac{d}{dt}w_k(t)$$

$$+\sum_{j=0}^{P}\sum_{k=0}^{P} K_j C_{0jk} w_k(t) = f(t), \text{ and} \tag{8.12}$$

$$\sum_{j=0}^{P}\sum_{k=0}^{P} M_j C_{ijk}\frac{d^2}{dt^2}w_k(t) + \sum_{j=0}^{P}\sum_{k=0}^{P} D_j C_{ijk}\frac{d}{dt}w_k(t)$$

$$+\sum_{j=0}^{P}\sum_{k=0}^{P} K_j C_{ijk} w_k(t) = 0\ \forall i = 1, 2, \cdots, P. \tag{8.13}$$

Table 8.3 C_{ijk} for Hermite polynomials for $P = 4$

C_{ijk}	Value	C_{ijk}	Value	C_{ijk}	Value
C_{000}	1	C_{112}	2	C_{233}	36
C_{011}	1	C_{123}	6	C_{244}	192
C_{022}	2	C_{134}	24	C_{334}	216
C_{033}	6	C_{222}	8	C_{444}	1728
C_{044}	24	C_{224}	24	$o.w.$	0

We use Tables 8.2 and 8.3 to expand Eqs. (8.12) and (8.13). We get $P + 1$ equations with $4(P + 1)$ variables and hence will require a sampling strategy to solve the system. Solving the above ODEs for $w_i(t)$ allows us to find the stochastic representation of the solution:

$$w(t; \xi) = \sum_{i=0}^{4} w_i(t) H_i(\xi).$$

We note that $w_0(t)$ is the total solution of the nominal case of the stochastic differential equation. From Eq. (8.5) we know that the mean of $w(t; \xi)$ is indeed $w_0(t)$. We also note that each of w_i for $i \neq 0$ represents the homogenous solution of the system.

8.3.3 *PCE with Galerkin Projection Applied to the Tapered PEH Model*

More specifically, we prepare the differential algebraic equation in Eq. (8.2) for the single-mode tapered PEH model reproduced as

$$\frac{d\mathbf{q}(t, \boldsymbol{\xi})}{dt} = \boldsymbol{A}(\boldsymbol{\xi})\mathbf{q}(t, \boldsymbol{\xi}) + \boldsymbol{B}(\boldsymbol{\xi})\mathbf{u}(t),$$

where

$$\mathbf{q}(t, \boldsymbol{\xi}) = \begin{bmatrix} \dot{\eta}(t, \boldsymbol{\xi}) \\ \eta(t, \boldsymbol{\xi}) \\ v(t, \boldsymbol{\xi}) \end{bmatrix},$$

$$\mathbf{A}_r(\boldsymbol{\xi}) = \begin{bmatrix} -2\zeta_r^{\xi}\omega_r^{\xi} & -(\omega_r^{\xi})^2 & \chi_r^{\xi} \\ 1 & 0 & 0 \\ \frac{\chi_r^{\xi}}{C_p^{\xi}} & 0 & -\frac{1}{R_l C_p^{\xi}} \end{bmatrix}, \text{ and}$$

$$\mathbf{B}_r(\boldsymbol{\xi}) = \begin{bmatrix} -P_0^m(\phi_r(\boldsymbol{\xi})) & -P_1^m(\phi_r(\boldsymbol{\xi})) \\ 0 & 0 \\ 0 & 0 \end{bmatrix}.$$

We apply the PCE of the state vector using the relationships

$$\hat{\eta}(t,\xi) = \sum_{k=0}^{P} \eta_k(t) H_k(\xi) \text{ and}$$

$$\hat{v}(t,\xi) = \sum_{k=0}^{P} v_k(t) H_k(\xi).$$

We also apply PCE on the system parameter matrices $\mathbf{A}(\xi)$ and $\mathbf{B}(\xi)$ as

$$\mathbf{A}(\xi) \approx \hat{\mathbf{A}}(\xi) = \sum_{k=0}^{P} \mathbf{A}_k H_k(\xi) \text{and}$$

$$\mathbf{B}(\xi) \approx \hat{\mathbf{B}}(\xi) = \sum_{k=0}^{P} \mathbf{B}_k H_k(\xi).$$

Here the coefficients can be obtained using the classical projection theorem as

$$\mathbf{A}_k = \frac{\langle \mathbf{A}(\xi), H_k(\xi) \rangle}{\langle H_k(\xi), H_k(\xi) \rangle} \text{and}$$
$$\mathbf{B}_k = \frac{\langle \mathbf{B}(\xi), H_k(\xi) \rangle}{\langle H_k(\xi), H_k(\xi) \rangle}. \tag{8.14}$$

Substituting in Eq. (8.2) we get the residual function as

$$\frac{d\hat{\mathbf{q}}(t,\xi)}{dt} = \hat{A}(\xi)\hat{\mathbf{q}}(t,\xi) + \hat{B}(\xi)\mathbf{u}(t).$$

$$\Longrightarrow \frac{d}{dt}\begin{bmatrix} \dot{\hat{\eta}}(t,\xi) \\ \hat{\eta}(t,\xi) \\ \hat{v}(t,\xi) \end{bmatrix} = \sum_{k=0}^{P} \mathbf{A}_k H_k(\xi) \begin{bmatrix} \dot{\hat{\eta}}(t,\xi) \\ \hat{\eta}(t,\xi) \\ \hat{v}(t,\xi) \end{bmatrix} + \sum_{k=0}^{P} \mathbf{B}_k H_k(\xi)\mathbf{u}(t).$$

$$\Longrightarrow \frac{d}{dt}\sum_{k=1}^{P} \begin{bmatrix} \dot{\eta}_k(t) \\ \eta_k(t) \\ v_k(t) \end{bmatrix} H_k(\xi) = \sum_{j=0}^{P}\sum_{k=0}^{P} \mathbf{A}_k H_k(\xi) \begin{bmatrix} \dot{\eta}_j(t) \\ \eta_j(t) \\ v_j(t) \end{bmatrix} H_j(\xi)$$
$$+ \sum_{k=0}^{P} \mathbf{B}_k H_k(\xi)\mathbf{u}(t)\ .$$

We now apply the Galerkin projection as

$$\frac{d}{dt}\left\langle \sum_{k=0}^{P} \begin{bmatrix} \dot{\eta}_k(t) \\ \eta_k(t) \\ v_k(t) \end{bmatrix} H_k(\boldsymbol{\xi}),\ H_i(\boldsymbol{\xi}) \right\rangle$$

$$= \left\langle \sum_{j=0}^{P} \sum_{k=0}^{P} \mathbf{A}_k H_k(\boldsymbol{\xi}) \begin{bmatrix} \dot{\eta}_j(t) \\ \eta_j(t) \\ v_j(t) \end{bmatrix} H_j(\boldsymbol{\xi}),\ H_i(\boldsymbol{\xi}) \right\rangle + \left\langle \sum_{k=0}^{P} \mathbf{B}_k H_k(\boldsymbol{\xi}),\ H_i(\boldsymbol{\xi}) \right\rangle \mathbf{u}(t).$$

Using the linearity property of the inner product we have

$$\frac{d}{dt} \sum_{k=0}^{P} \begin{bmatrix} \dot{\eta}_k(t) \\ \eta_k(t) \\ v_k(t) \end{bmatrix} \langle H_k(\boldsymbol{\xi}),\ H_i(\boldsymbol{\xi}) \rangle = \sum_{j=0}^{P} \sum_{k=0}^{P} \mathbf{A}_k \begin{bmatrix} \dot{\eta}_j(t) \\ \eta_j(t) \\ v_j(t) \end{bmatrix} \left\langle H_k(\boldsymbol{\xi}) H_j(\boldsymbol{\xi}),\ H_i(\boldsymbol{\xi}) \right\rangle$$

$$+ \sum_{k=0}^{P} \mathbf{B}_k \langle H_k(\boldsymbol{\xi}),\ H_i(\boldsymbol{\xi}) \rangle \mathbf{u}(t).$$

Using the orthogonality property of the gPC basis functions, the RHS is simplified to

$$\frac{d}{dt} \sum_{k=0}^{P} \begin{bmatrix} \dot{\eta}_k(t) \\ \eta_k(t) \\ v_k(t) \end{bmatrix} \underbrace{\langle H_k(\boldsymbol{\xi}),\ H_i(\boldsymbol{\xi}) \rangle}_{=0,\ \forall k \neq i} = \frac{d}{dt} \begin{bmatrix} \dot{\eta}_i(t) \\ \eta_i(t) \\ v_i(t) \end{bmatrix} \langle H_i(\boldsymbol{\xi}),\ H_i(\boldsymbol{\xi}) \rangle,$$

whereas the LHS is simplified to

$$\sum_{j=0}^{P} \sum_{k=0}^{P} \mathbf{A}_k \begin{bmatrix} \dot{\eta}_j(t) \\ \eta_j(t) \\ v_j(t) \end{bmatrix} \left\langle H_k(\boldsymbol{\xi}) H_j(\boldsymbol{\xi}),\ H_i(\boldsymbol{\xi}) \right\rangle + \sum_{k=0}^{P} \mathbf{B}_k \underbrace{\langle H_k(\boldsymbol{\xi}),\ H_i(\boldsymbol{\xi}) \rangle}_{=0,\ \forall k \neq i} \mathbf{u}(t) =$$

$$\sum_{j=0}^{P} \sum_{k=0}^{P} \mathbf{A}_k \begin{bmatrix} \dot{\eta}_j(t) \\ \eta_j(t) \\ v_j(t) \end{bmatrix} \left\langle H_k(\boldsymbol{\xi}) H_j(\boldsymbol{\xi}),\ H_i(\boldsymbol{\xi}) \right\rangle + \mathbf{B}_i \langle H_i(\boldsymbol{\xi}),\ H_i(\boldsymbol{\xi}) \rangle \mathbf{u}(t).$$

The DAE after simplification becomes

$$\frac{d}{dt} \begin{bmatrix} \dot{\eta}_i(t) \\ \eta_i(t) \\ v_i(t) \end{bmatrix} \langle H_i(\boldsymbol{\xi}),\ H_i(\boldsymbol{\xi}) \rangle = \sum_{j=0}^{P} \sum_{k=0}^{P} \mathbf{A}_k \begin{bmatrix} \dot{\eta}_j(t) \\ \eta_j(t) \\ v_j(t) \end{bmatrix} \left\langle H_k(\boldsymbol{\xi}) H_j(\boldsymbol{\xi}),\ H_i(\boldsymbol{\xi}) \right\rangle$$

$$+ \mathbf{B}_i \langle H_i(\boldsymbol{\xi}),\ H_i(\boldsymbol{\xi}) \rangle \mathbf{u}(t).$$

We divide out the inner product term $\langle H_i(\boldsymbol{\xi}),\ H_i(\boldsymbol{\xi})\rangle$ on both sides to get

$$\frac{d}{dt}\begin{bmatrix}\dot{\eta}_i(t)\\ \eta_i(t)\\ v_i(t)\end{bmatrix} = \sum_{j=0}^{P}\sum_{k=0}^{P}\mathbf{A}_k\begin{bmatrix}\dot{\eta}_j(t)\\ \eta_j(t)\\ v_j(t)\end{bmatrix}\frac{\left\langle H_k(\boldsymbol{\xi})H_j(\boldsymbol{\xi}),\ H_i(\boldsymbol{\xi})\right\rangle}{\langle H_i(\boldsymbol{\xi}),\ H_i(\boldsymbol{\xi})\rangle} + \mathbf{B}_i\mathbf{u}(t). \tag{8.15}$$

Equation (8.15) represents $P + 1$ systems of three equations each. We define

$$\alpha_{kji} = \frac{\left\langle H_k(\boldsymbol{\xi})H_j(\boldsymbol{\xi}),\ H_i(\boldsymbol{\xi})\right\rangle}{\langle H_i(\boldsymbol{\xi}),\ H_i(\boldsymbol{\xi})\rangle}.$$

The equation is now recast by converting the summation over k in the LHS to a matrix multiplication as

$$\frac{d}{dt}\begin{pmatrix}\dot{\eta}_i(t)\\ \eta_i(t)\\ v_i(t)\end{pmatrix} = \left[\cdots\ \sum_{k=1}^{P}\mathbf{A}_k\alpha_{kji}\ \cdots\right]\begin{bmatrix}\vdots\\ \dot{\eta}_j(t)\\ \eta_j(t)\\ v_j(t)\\ \vdots\end{bmatrix} + \mathbf{B}_i\mathbf{u}(t).$$

Defining $\tilde{\boldsymbol{q}}(t) = [\dot{\eta}_0(t), \eta_0(t), v_0(t), \cdots, \dot{\eta}_P(t), \eta_P(t), v_P(t)]^T$, and $\tilde{\mathbf{B}} = \begin{bmatrix}\vdots\\ \mathbf{B}_i\\ \vdots\end{bmatrix}$, and $\tilde{\mathbf{A}}$ such that $\tilde{\mathbf{A}}_{ij} = \sum_{k=0}^{P}\mathbf{A}_k\alpha_{kji}$, the final state derivative equation after the Galerkin projection becomes

$$\frac{d\tilde{\boldsymbol{q}}(t)}{dt} = \tilde{\boldsymbol{A}}\tilde{\boldsymbol{q}}(t) + \tilde{\boldsymbol{B}}\mathbf{u}(t). \tag{8.16}$$

The algorithm to complete the PCE Galerkin projection-based UQ is given in Algorithm inset 8.1.

8.3.3.1 Augmented Model Interpretation

The stochastic DAE obtained after polynomial chaos expansion has been interpreted by Manfredi [7, 8] as an augmented circuit with sample points as additional circuit components. This higher-order circuit can be solved using a circuit solver. We notice in Eq. (8.16) that after Galerkin projection of the PCE we get an augmented system defined by matrices $\tilde{\boldsymbol{A}}$ and $\tilde{\boldsymbol{B}}$ and augmented state vector $\tilde{\boldsymbol{q}}(t)$ with $3 \times P$ states. To solve for the augmented states, we can use the same solver as before but for a larger system.

Algorithm 8.1 UQ using Galerkin projection on PCE of a tapered beam EH

procedure PCUQOFTPEH($\mathbf{A}$, $\mathbf{B}$, $\mathbf{u}(t)$)

Prepare ensemble space

Identify independent design parameters with variability impacting device performance:

$$\boldsymbol{\xi} = \left[\xi^1, \xi^2, \ldots, \xi^n\right]^T \text{ and } \boldsymbol{\xi} \in \Omega.$$

Decide upon the PDF of $\boldsymbol{\xi}$, which assuming independence, is $p(\boldsymbol{\xi}) = \prod_{i=1}^{n} p(\xi_i)$.

Select the polynomial class to use for PCE: e.g. Hermite polynomials for Gaussian $p(\boldsymbol{\xi})$

Select the polynomial expansion order $P = \frac{(m+n)!}{m!n!}$ where m is highest total polynomial degree and n is the number of random variables.

Compute augmented system parameters

Compute $\mathbf{A}_k$, *and* $\mathbf{B}_k$ using projection as in Eq. (8.14). Use the Gauss-quadrature rule for the integrals.

Build $\tilde{\mathbf{A}}$ such that $\tilde{\mathbf{A}}_{ij} = \sum_{k=0}^{P} \mathbf{A}_k \alpha_{kji}$

Build $\tilde{\mathbf{B}} = \left[\cdots \mathbf{B}_i^T \cdots\right]^T$

Solve augmented system

Solve the augmented system equation $\frac{d\tilde{q}(t)}{dt} = \tilde{\boldsymbol{A}}\tilde{\boldsymbol{q}}(t) + \tilde{\boldsymbol{B}}\mathbf{u}(t)$ for the augmented state vector $\tilde{\boldsymbol{q}}(t) = [\dot{\eta}_0(t), \eta_0(t).v_0(t), \cdots, \dot{\eta}_P(t), \eta_P(t).v_P(t)]^T$

Compute Moments

Compute the required statistics as

$$mean\,[\eta(t)] = \eta_0(t)\alpha_0,$$

$$mean\,[v(t)] = v_0(t)\alpha_0,$$

$$var\,[\eta(t)] = \sum_{k=1}^{P} \eta_k^2(t)\alpha_k^2, \text{ and}$$

$$var\,[v(t)] = \sum_{k=1}^{P} v_k^2(t)\alpha_k^2.$$

where $\alpha_k^2 = \langle H_k, H_k\rangle$.

end procedure

8.4 Stochastic Testing

Stochastic testing is a generalized polynomial chaos-based intrusive simulator for uncertainty quantification of a transistor-level circuit simulation. This simulator is a variant of the interpolation-based stochastic collocation and sets up the differential algebraic equation (DAE) and solves it at a subset of quadrature points.

8.4.1 Setting Up the DAE

We start by considering a general non-linear circuit for which the DAE can be written as

$$\frac{d}{dt}\mathbf{q}(\mathbf{x}(t,\boldsymbol{\xi}),\boldsymbol{\xi}) + \mathbf{f}(\mathbf{x}(t,\boldsymbol{\xi}),\boldsymbol{\xi}) = B\mathbf{u}(t), \tag{8.17}$$

where the state vector $\mathbf{x} \in \mathbb{R}^n$ is a vector of the nodal voltages, $\mathbf{u} \in \mathbb{R}^m$ is the input, $\mathbf{q} \in \mathbb{R}^n$ represents the charge flux term, and $\mathbf{f} \in \mathbb{R}^n$ is the current or voltage term. The vector $\boldsymbol{\xi} = [\xi_1, \cdots, \xi_l]$ represents the low-level random variables that describe the device-level uncertainties and are assumed mutually independent. The purpose of a stochastic solver is to extract statistical information of the state vector $\mathbf{x}(t,\boldsymbol{\xi})$ such as different moments and the PDF from the stochastic DAE. Invoking PCE we estimate the state vector using analysis in the gPC bases domain as

$$\hat{x}(t,\boldsymbol{\xi}) = \sum_{k=1}^{P} \hat{x}_k(t) H_k(\boldsymbol{\xi}),$$

where $H_k(\boldsymbol{\xi})$ is the kth gPC basis function. Substituting in the DAE equation (Eq. 8.17) we get the residual function as

$$Res(\boldsymbol{X}(t),\boldsymbol{\xi}) = \frac{d}{dt}\mathbf{q}(\hat{x}(t,\boldsymbol{\xi}),\boldsymbol{\xi}) + \mathbf{f}(\hat{x}(t,\boldsymbol{\xi}),\boldsymbol{\xi}) - B\mathbf{u}(t), \tag{8.18}$$

where the new unknown state vector is $\boldsymbol{X}(t) = \left[\hat{x}_1(t); \cdots; \hat{x}_P(t)\right] \in \mathbb{R}^{nP}$.

Stochastic testing starts from Eq. (8.18) and enforces the residual function to be zero at each of the P selected testing points $\boldsymbol{\xi}^1, \cdots, \boldsymbol{\xi}^P$ which leads to the DAE [9]:

$$\frac{dQ(\boldsymbol{X}(t))}{dt} + F(\boldsymbol{X}(t)) = \tilde{B}\mathbf{u}(t), \tag{8.19}$$

where

$$Q(\boldsymbol{X}(t)) = \begin{bmatrix} \mathbf{q}(\hat{x}(t,\boldsymbol{\xi}^1),\boldsymbol{\xi}^1) \\ \vdots \\ \mathbf{q}(\hat{x}(t,\boldsymbol{\xi}^P),\boldsymbol{\xi}^P) \end{bmatrix},$$

$$F(\boldsymbol{X}(t)) = \begin{bmatrix} \mathbf{f}(\hat{x}(t,\boldsymbol{\xi}^1),\boldsymbol{\xi}^1) \\ \vdots \\ \mathbf{f}(\hat{x}(t,\boldsymbol{\xi}^P),\boldsymbol{\xi}^P) \end{bmatrix}, \text{ and}$$

$$\tilde{B} = \begin{bmatrix} B \\ \vdots \\ B \end{bmatrix}.$$

Stochastic testing directly solves the DAE in Eq. (8.19) using a specialized intrusive transient solver to compute the gPC coefficients $\left\{\hat{x}_k\right\}_{k=1}^{P}$.

8.4.2 *Testing Nodes*

The testing nodes are selected for stochastic testing by first generating $(m+1)^l \gg P$ candidate nodes using the Gaussian-quadrature rule and then selecting the statistically important candidate nodes with higher weight values. Let $\xi_k \in \Omega_k$ be a random device-level parameter and $p_k(\xi_k)$ its PDF. The Gaussian quadrature can be used to estimate a one-dimensional stochastic integral as

$$\int_{\Omega_k} g(\xi_k) p_k(\xi_k) d\xi_k \approx \sum_{j=1}^{\hat{n}} g(\xi_k^j) w_k^j, \tag{8.20}$$

where ξ_k^j is the jth quadrature point of the kth parameter and w_k^j is the corresponding weight. A multivariate stochastic integral is evaluated similarly as

$$\int_{\Omega_k} g(\boldsymbol{\xi}) p(\boldsymbol{\xi}) d\boldsymbol{\xi} \approx \sum_{j=1}^{\hat{n}^l} g(\boldsymbol{\xi}^j) w^j.$$

8.5 Uncertainty Propagation of the Tapered Beam Model

While a PCE-based stochastic solver is needed to solve the dynamic governing system of the tapered beam model, many of the parameters used by the model are higher-level derived quantities that depend upon process variabilities through various functionals.

8.5.1 *PCE of the Tapered Beam Coupled Electromechanical Model*

The stochastic electromechanical state-space system described in Sect. 8.1.2 depends on the uncertain system parameters ζ_r^ξ, ω_r^ξ, χ_r^ξ, C_p^ξ, and $F_r(t, \boldsymbol{\xi})$. These parameters depend on aleatoric uncertainty in $\boldsymbol{\xi}$ through the relationships

$$\omega_r^\xi = \sqrt{\frac{Y I_0^\xi \beta_r}{m_0^\xi}},$$

$$\zeta_r^\xi = \frac{c_s I_0^\xi \omega_r^\xi}{2YI_0^\xi} + \frac{c_a P_0(\phi_r^2(\boldsymbol{\xi}))}{2\omega_r^\xi},$$

$$C_p^\xi = \frac{\varepsilon_{33}^S b_0 l}{h_p^\xi}(1 - \frac{\alpha}{2}),$$

$$\chi_r^\xi = -Y_p d_{31} h_{pc}^\xi b_0 \left((1-\alpha)\phi_r'(l, \boldsymbol{\xi}) + \frac{\alpha}{l}\phi_r(l, \boldsymbol{\xi})\right), \text{ and}$$

$$f_{g,r}(t, \boldsymbol{\xi}) = P_0^{(m)}(\phi_r(\boldsymbol{\xi})),$$

where $P_0^{(m)}(\cdot)$ is the mass-weighted inner product defined in Eq. (6.10). Let the quantities on the LHS for mode r form a set

$$\Gamma_r^\xi = \left\{\omega_r^\xi, \zeta_r^\xi, C_p^\xi, \chi_r^\xi, f_{g,r}(t, \boldsymbol{\xi})\right\}.$$

8.5.2 Probability Transformation

When the random variable y depends on a single random variable x with mapping f such that $y = f(x)$, the PDF of y, $p_Y(y)$ can be found using the transformation

$$p_Y(y) = \sum_{x, f(x)=y} \frac{p_X(x)}{|f'(x)|}, \tag{8.21}$$

which can also be written as

$$p_Y(y) = \int_{\mathbb{R}} \delta(y - f(x)) p_X(x) dx = \langle \delta(y - f(x)) \rangle,$$

When the higher-level random variable inherits its uncertainty from multiple independent random variables $\{x_i\}_{i=1}^n$, the resulting PDF of y can be found using the formula:

$$p_Y(y) = \int_{\mathbb{R}^{n-1}} \sum_{x_1, f(\mathbf{x})=y} \frac{p_X(\mathbf{x})}{\left|\frac{\partial f(\mathbf{x})}{\partial x_1}\right|} dx_2 \ldots dx_n.$$

8.5.3 Applying Probability Transformation to Uncertainty Propagation

Now taking each element from $\Gamma_r(\boldsymbol{\xi})$, we can attempt to find its PDF using Eq. (8.21). To simplify the analysis we include only the thickness of the PZT layer as a random variable with a standard deviation of a third of its tolerance, which for Yang's tapered beam structure used in the example in Sect. 6.6 is assumed to be 10% of its nominal design value. The mean of the distribution is the nominal design thickness value. Hence, the PDF of the PZT thickness is $h_p \sim \mathcal{N}(\mu_{h_p}, \sigma_{h_p}^2)$ where $\mu_{h_p} = 5e - 4$ and $\sigma_{h_p}^2 = 0.0167$. Dropping the p subscript, the PDF of the PZT thickness is then expressed as the unidimensional normal distribution $p_H(h) = \frac{1}{\sqrt{2\pi\sigma_h^2}} e^{-\frac{(h-\mu_h)^2}{\sigma^2}}$. We can systematically compute the PDFs of each of the uncertain system parameters by carefully setting up their expressions in terms of h_p and analytically solving Eq. (8.21) for $p_Y(y)$. Alternatively, we can numerically compute the same by generating a vector of h_p values within the tolerance specified and computing corresponding vectors for each one of the system parameters numerically. We can then use Eq. (8.21) to compute the PDFs of these parameters numerically.

8.5.4 Uncertainty Dependency Graph

Figure 8.1 shows the uncertainty dependency graph for the tapered PEH model. This graph can be used to build the uncertainty model for the ECM variability incrementally from process variability. The orange nodes are the parameters of the device design with aleatoric uncertainty which are modeled with a PDF as above. The yellow nodes are the intermediate-level device parameters that capture some of the propagated system uncertainties. The blue nodes are system parameters that appear in the state-space system. The green nodes are parameters of the equivalent circuit model and the magenta nodes are the steady-state system outputs that do not need a solver to be evaluated.

8.5.5 Uncertainty Propagation Using Probability Transformation

For illustration purposes, we consider the uncertainty in only the piezoelectric layer thickness and propagate it forward to system parameters. The dependency graph simplifies when input uncertainties are limited to only one parameter as shown in Fig. 8.2.

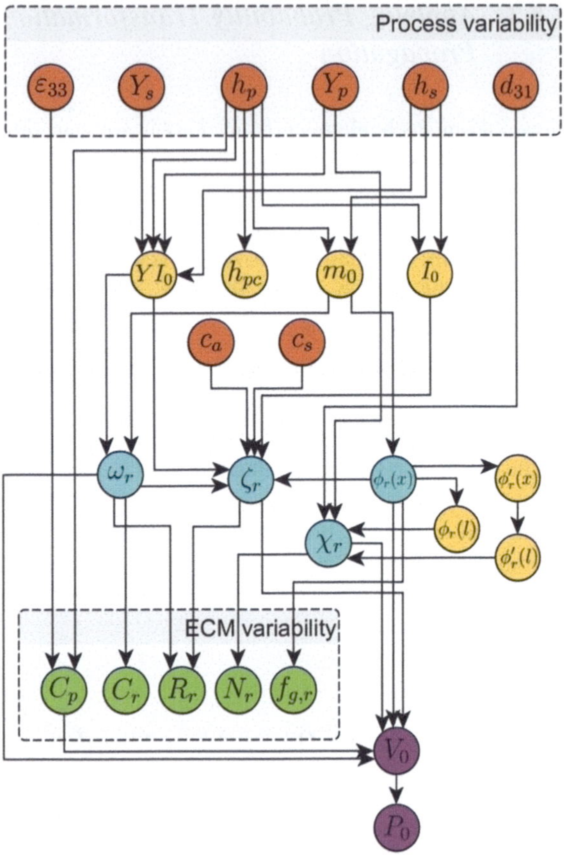

Fig. 8.1 Uncertainty dependency graph for the tapered beam EH model

We start with a probability model for the piezoelectric layer thickness, which as described above is considered to be a Gaussian distribution with a standard deviation equal to a third of the process tolerance for piezoelectric layer thickness such that

$$p_H(h) = \frac{1}{\sqrt{2\pi\sigma_h^2}} e^{-\frac{(h-\mu_h)^2}{\sigma^2}}, \tag{8.22}$$

where $\mu_h = 5e - 4$ and $\sigma_h^2 = 0.0167$. We can set up a dense h vector $[h_1, h_2, \cdots, h_{N_s}]$ and evaluate each of the system parameters corresponding to each of the $\{h_i\}_{i=1}^{N_s}$. We then process the resulting vectors using the PDF transformation equation (Eq. 8.21) to compute the PDF of the system parameters. Note that the derivative in Eq. (8.21) is computed numerically.

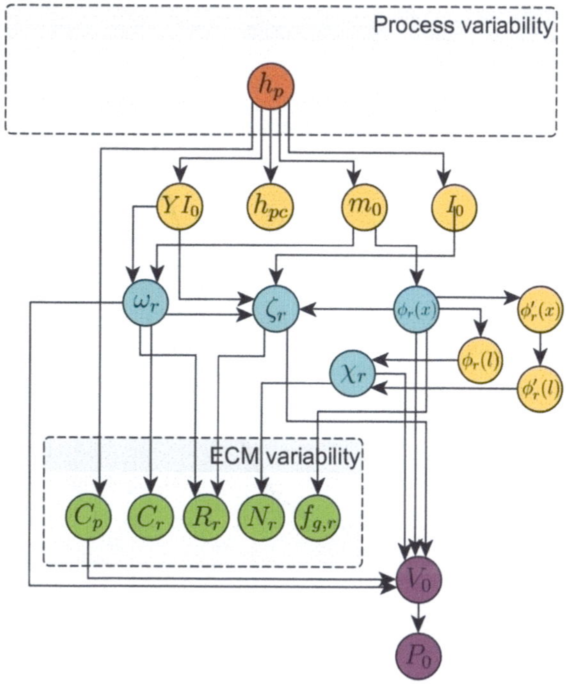

Fig. 8.2 Uncertainty dependency graph for the tapered beam EH model with a single aleatoric uncertainty

8.5.5.1 PDF Transformation Results

We used the Monte Carlo simulation results in Chap. 7 to create PDF models for the various quantities using kernel density estimation. We also used Eq. (8.21) to propagate the uncertainty to the desired node on the uncertainty dependency graph in Fig. 8.2. We start by looking at the PDF of the PZT layer thickness generated from the normal distribution shown in Eq. (8.22). The kernel density estimation is shown in dashed red and the PDF estimation using probability transformation of Eq. (8.21) is shown in blue (Fig. 8.3).

Next, we propagate the uncertainty to the intermediate parameters shown in orange in the uncertainty propagation graph in Fig. 8.2. We show in Fig. 8.4 the PDFs of the mass density at the base m_0, the effective Young's modulus Y_0 at the base, and the homogenized flexural rigidity at the base of the cantilever YI_0.

We propagate the uncertainty further down to obtain the PDFs of the ECM parameters. PDFs of C_p, R_r, C_r, and N_r are shown in Fig. 8.5.

Finally, we propagate the uncertainties through the model to the steady-state outputs, including the optimum impedance value. We show these results in Fig. 8.6.

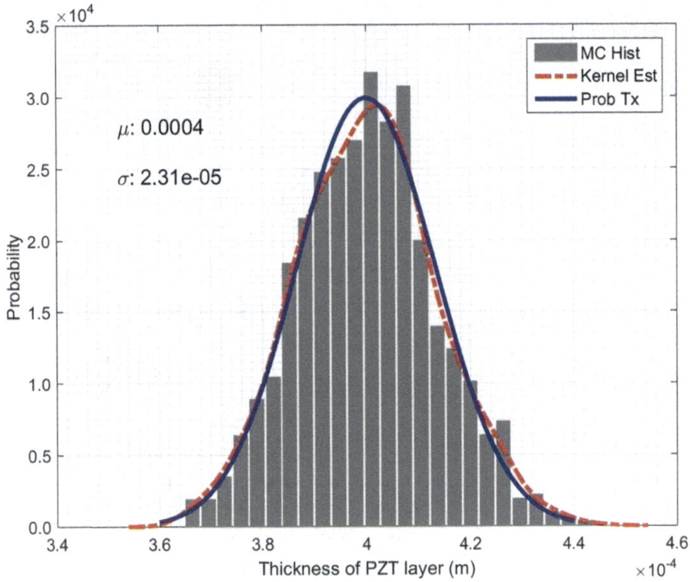

Fig. 8.3 PDF of PZT layer thickness. The data is generated from a normal distribution

8.6 Hierarchical Uncertainty Quantification of Tapered PEH

We use the stochastic testing simulator developed by Zhang et al. [9] to solve the ECM of the tapered beam EH once the uncertainty in the ECM has been evaluated using MC or other non-intrusive uncertainty propagation strategies. The hierarchical UQ used by Zhang [1] introduces q surrogate models that describe system-level parameters $\{\hat{x}_i\}_{i=1}^{q}$ each depending on a set of lower-level uncertain parameters $\{\boldsymbol{\xi}_i\}_{i=1}^{q}$ through a function f_i as

$$\hat{x}_i = f_i(\boldsymbol{\xi}_i),\ with\, \boldsymbol{\xi}_i \in \mathbb{R}^{d_i}\ i = 1, \cdots, q.$$

Figure 8.7 shows the flowchart for performing uncertainty quantification in the EH model of a tapered beam hierarchically. The technique includes the preparation of the *surrogate* models before solving the DAE using stochastic testing. Here lower-level uncertain parameters $\{\boldsymbol{\xi}_i\}_{i=1}^{q}$ include the process variability sources shown as the orange nodes and the system-level parameters $\{\hat{x}_i\}_{i=1}^{q}$ are the ECM parameters shown as the green nodes in Figs. 8.1 and 8.2. In the first stage, the PDFs for the ECM parameters are evaluated using strategies discussed in Sect. 8.5. As an alternative, we can also propagate the moments of the distribution, assume a probability model for the ECM parameters, and provide the estimated moments to this model. However, the methodology that we follow in this work uses Monte

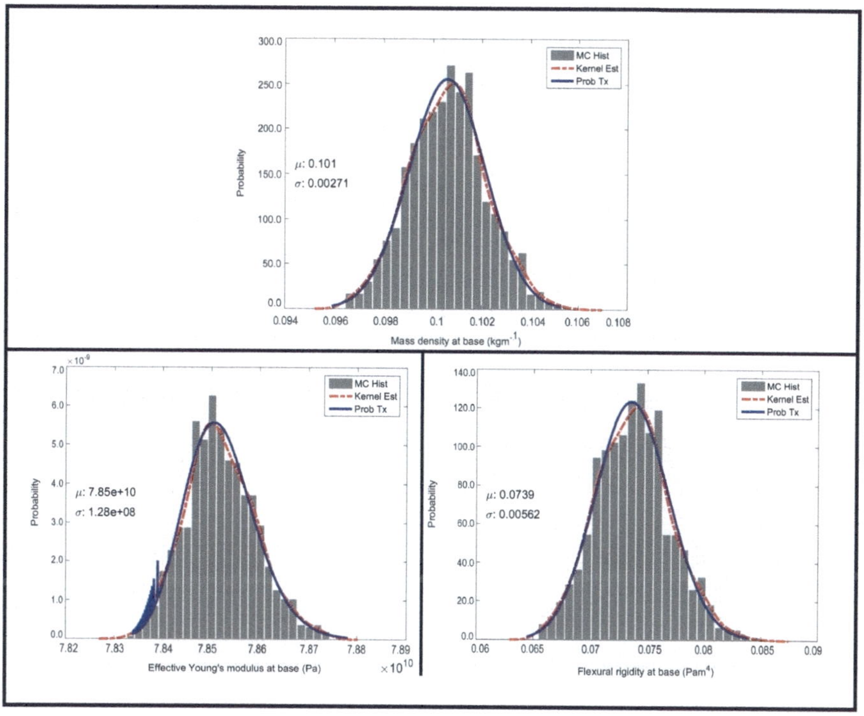

Fig. 8.4 PDF of intermediate parameters

Carlo simulation to estimate the PDFs of the high-level ECM parameters. The stochastic circuit equation can then be solved using the stochastic testing algorithm [9] explained in Sect. 8.4.

8.6.1 Estimating the PDFs of ECM Parameters

The PDFs of the ECM parameters can be estimated using the following steps [1]:

1. Use Monte Carlo simulation over N_s samples to obtain the discrete cumulative density function (CDF) curve of $\hat{x}_i = f_i(\boldsymbol{\xi}_i)$.
2. For each i, perform a linear transformation on $\hat{x}_i$ to scale its range to $(0, 1)$. The scaled parameter is called x_i.
3. Pick $n_s \ll N_s$ points (x_i, y_i) from the CDF of x.
4. Use a monotone interpolation method to interpolate the n_s-point CDF above to obtain a smooth monotone function $p_m(x)$ which is the CDF of x.
5. Differentiate $p_m(x)$ to find the closed-form expression for the PDF, $p(x)$.

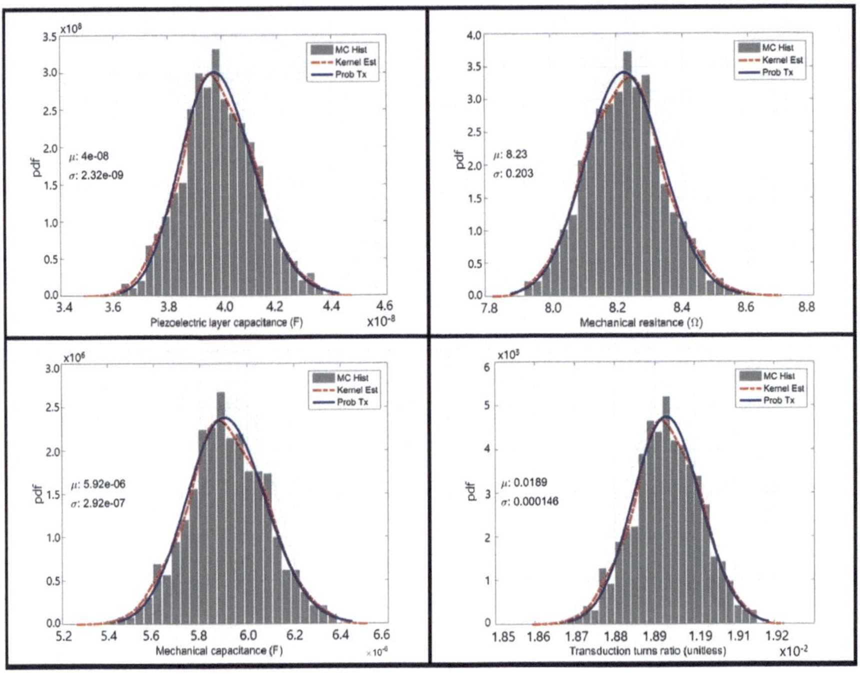

Fig. 8.5 PDFs of ECM parameters

8.6.2 Determining the gPC Basis Functions

Once the PDF of the ECM parameters is known, the orthonormal gPC basis functions are computed using the methodology explained in Sect. 8.2.1. The basis functions $H_i(x)$ are computed using the recurrence relation:

$$\pi_{i+1}(x) = (x - \gamma_i)\pi_i(x) - \kappa_i \pi_{i-1}(x) \ \forall i = 0, 1, \cdots, \hat{n},$$

with the initial conditions for the recursion that $\pi_{-1}(x) = 0$ and $\pi_0(x) = 1$. The expressions for γ_i and κ_{i+1} are elaborated using the definitions of the expected value as

$$\gamma_i = \frac{\int_{\mathbb{R}} x\pi_i^2(x)p(x)dx}{\int_{\mathbb{R}} \pi_i^2(x)p(x)dx}, \ \kappa_{i+1} = \frac{\int_{\mathbb{R}} \pi_{i+1}^2(x)p(x)dx}{\int_{\mathbb{R}} \pi_i^2(x)p(x)dx} \ \forall i = 0, 1, \cdots, \hat{n}.$$

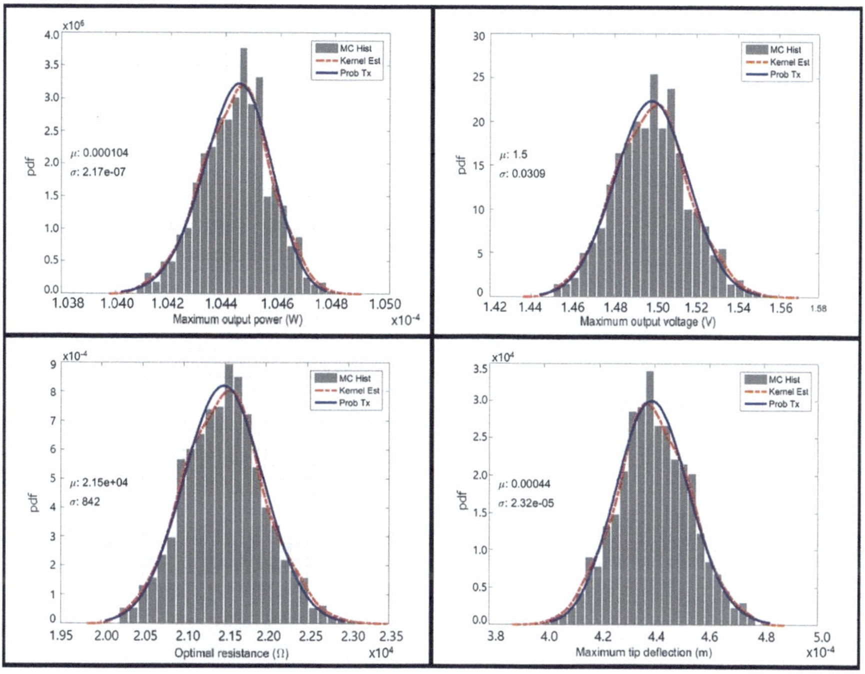

Fig. 8.6 PDFs of steady-state outputs

The basis functions are then computed as

$$H_i(x) = \frac{\pi_i(x)}{\sqrt{\prod_{j=0}^{i} \kappa_j}} \quad \forall i = 0, 1, \cdots, \hat{n}.$$

8.6.3 Determining the Quadrature Points

Gaussian quadrature rule helps evaluate stochastic integrals shown in Eq. (8.20). To find the quadrature points ξ_k^j and the weights w_k^j, we first set up the symmetric tridiagonal matrix $\mathbf{J}$:

$$\mathbf{J}_k = \begin{bmatrix} \gamma_0^k & \sqrt{\kappa_1^k} & & & \\ \sqrt{\kappa_1^k} & \gamma_1^k & \ddots & & \\ & \ddots & \ddots & \ddots & \\ & & \ddots & \gamma_{\hat{n}-1}^k & \sqrt{\kappa_{\hat{n}}^k} \\ & & & \sqrt{\kappa_{\hat{n}}^k} & \gamma_{\hat{n}}^k \end{bmatrix}.$$

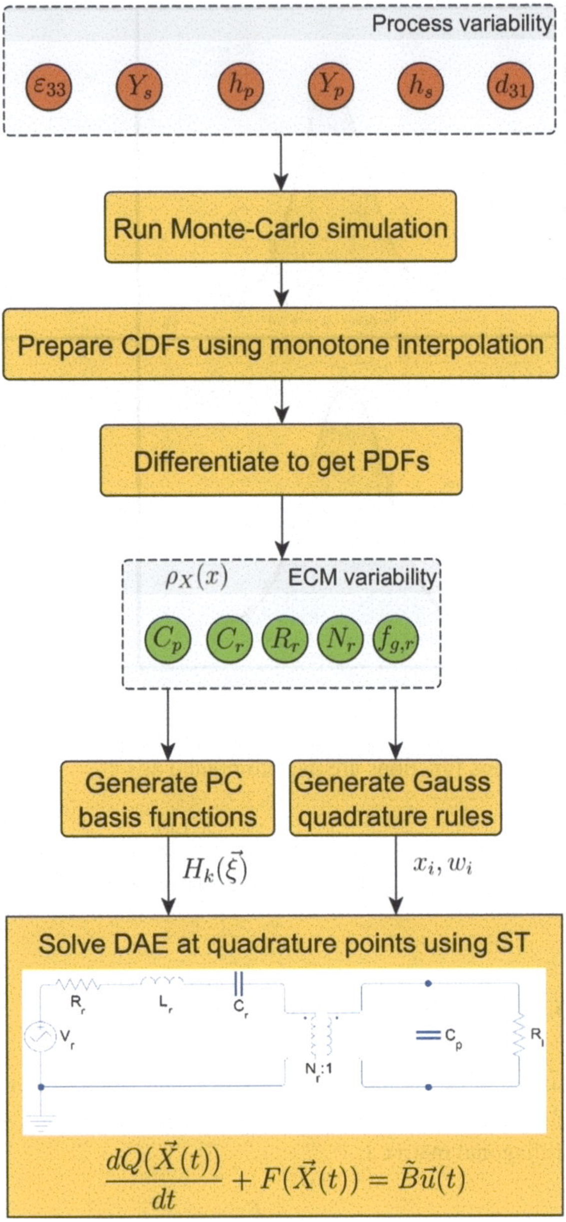

Fig. 8.7 Uncertainty quantification workflow for tapered beam EH

We then perform eigenvalue decomposition of $\mathbf{J}_k$ such that $\mathbf{J}_k = \mathbf{U}_k \boldsymbol{\Sigma}_k \mathbf{U}_k^T$, where $\mathbf{U}_k$ is a unitary matrix with $u_{i,j}$ its (i, j)th entry. The quadrature points ξ_k^j are the jth diagonal element of Σ_k and w_j^k is $u_{1,j}^2$ of $\mathbf{U}_k$. With the generated basis functions, the DAE is set up at the quadrature points before an intrusive transient solver is invoked to solve the stochastic DAE.

8.7 Conclusion

In this chapter, we have shown that we can set up the stochastic governing equation with the electromechanical model of the tapered PEH and apply the PCE and Galerkin projection to it. The resulting system can be interpreted as an augmented state-space system and solved using conventional solvers. We constructed an uncertainty dependency graph to help propagate uncertainty models from the process parameters to the ECM system parameters. Consequently, we have proposed to use the hierarchical algorithm suggested by Zhang [1] on the TPEH ECM model with Monte Carlo simulations or probability transformation-based uncertainty propagation for the surrogate models.

References

1. Zheng Zhang et al. "Calculation of generalized polynomial-chaos basis functions and Gauss quadrature rules in hierarchical uncertainty quantification". In: *IEEE Transactions on Computer-Aided Design of Integrated Circuits and Systems* 33.5 (2014), pp. 728–740.
2. Zheng Zhang. "Uncertainty Quantification for Integrated Circuits and Microelectromechanical Systems". PhD thesis. Citeseer, 2015.
3. Norbert Wiener. "The homogeneous chaos". In: *American Journal of Mathematics* 60.4 (1938), pp. 897–936.
4. Dongbin Xiu and George Em Karniadakis. "Modeling uncertainty in steady state diffusion problems via generalized polynomial chaos". In: *Computer methods in applied mechanics and engineering* 191.43 (2002), pp. 4927–4948.
5. Dongbin Xiu and George Em Karniadakis. "The Wiener–Askey polynomial chaos for stochastic differential equations". In: *SIAM journal on scientific computing* 24.2 (2002), pp. 619–644.
6. Walter Gautschi. "On generating orthogonal polynomials". In: *SIAM Journal on Scientific and Statistical Computing* 3.3 (1982), pp. 289–317.
7. Paolo Manfredi, Igor S Stievano, and Flavio G Canavero. "Alternative SPICE implementation of circuit uncertainties based on orthogonal polynomials". In: *Electrical Performance of Electronic Packaging and Systems (EPEPS), 2011 IEEE 20th Conference on*. IEEE. 2011, pp. 41–44.
8. Paolo Manfredi. "High-speed interconnect models with stochastic parameter variability". PhD thesis. Politecnico di Torino, 2013.
9. Zheng Zhang et al. "Stochastic testing method for transistor-level uncertainty quantification based on generalized polynomial chaos". In: *IEEE Transactions on Computer-Aided Design of Integrated Circuits and Systems* 32.10 (2013), pp. 1533–1545.

Chapter 9
Variation-Aware Design of Reliable Piezoelectric MEMS Energy Harvesters

In this chapter, we propose a process variation-aware MEMS device design flow where a fast device model is available. We start with explaining the process-aware MEMS design flow used during this work. We present our motivation for considering process variability in the design flow and proceed to present two variants of process variation- aware MEMS design flow, one with the stochastic simulation as part of the optimization loop and one where it is decoupled from the optimization loop.

9.1 Process-Aware MEMS Device Design

As the demand for MEMS sensor quality, fidelity, and reliability increases, MEMS design workflow, much like CMOS design, must make use of CAD tools and methodologies to help predict device performance for design optimization. Most of the existing computational tools for MEMS design take the traditional MEMS design perspective of one MEMS process for one MEMS device and one sensor product. While very capable of modeling the 3D multi-domain physics of a MEMS sensor, these tools still require a significant amount of manual tuning to adapt them to product specifications, sensor environments, and MEMS process changes.

MEMS designs are inherently fabrication process dependent where the layered structure and its thicknesses are determined from the technology process stack. This means that a design exercise in MEMS usually requires discovering a design layout for optimal performance constrained by the relevant process stack and the material properties from the foundry technology database. The design process of a MEMS device includes exploring different configurations of the process stack itself. The process-aware design methodology is depicted in Fig. 9.1.

The key figure of merit for piezoelectric MEMS EH is power generation per unit area. Process-aware performance optimization requires discovering the dimensions

W. U. Syed, I. (Abe) M. Elfadel, *Tapered Beams in MEMS*,
https://doi.org/10.1007/978-3-031-66391-8_9

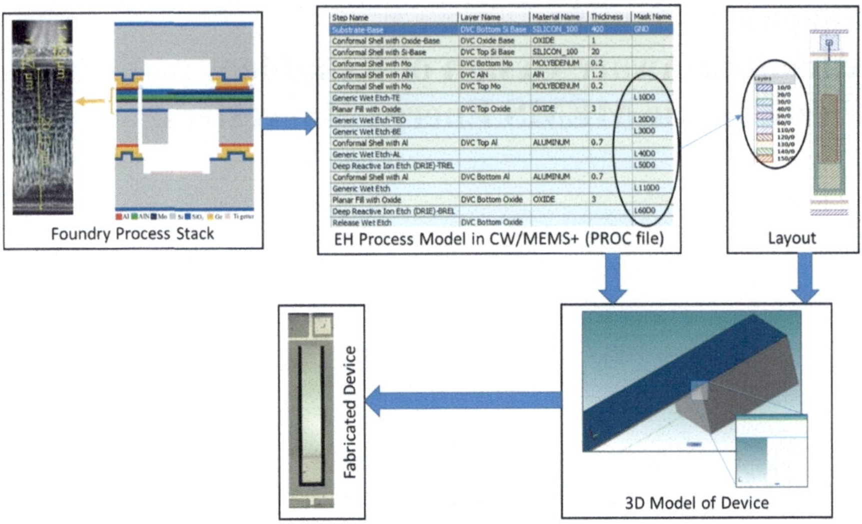

Fig. 9.1 Process-aware device modeling

of the device that would result in generating the maximum charge separation while optimizing electrode placement and polling configuration for maximum charge collection. We use an automated flow for the process-aware design of piezoelectric energy harvesters (PEHs) that includes multiple design components and uses multiple simulation and analysis tools and techniques. A parameterized MEMS+ model of the PEH is first constructed complete with shaped electrodes and squeezed film damping definition. MEMS+ is a network-based FEM solver software developed by Coventor Inc. that allows the construction of a process-aware device model. The foundry's technology information is encoded in the material (.MMDB) and process (.PROC) files. MEMS+ is used to simulate the device physics and compute the device output to electromechanical actuation, while MATLAB and Cadence are used to compute the power output across an optimal load. To speed up the simulation and design process we convert the parameterized MEMS+ model to its electrical circuit equivalent using the equivalent circuit model (ECM) presented in [1]. This equivalent circuit for each design realization is then co-simulated with the load circuit in Cadence or MATLAB in order to evaluate the output voltage and power. An optimization algorithm sweeps the design parameters to find the optimal design for maximal power density while imposing a list of design constraints. For the optimization of devices with doubly clamped beam structures or when optimizing structures for higher modes, an additional electrode placement optimization component is added to the design process. The layout of the final device design is exported from MEMS+ and is used to construct a CoventorWare model for detailed FEM simulation. CoventorWare, a detailed field solver-based FEM solver software also developed by Coventor Inc., is used to validate the

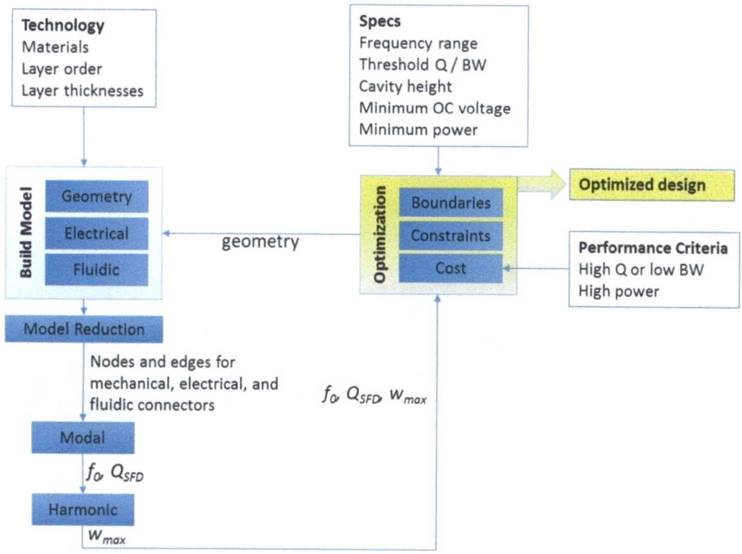

Fig. 9.2 Proposed flow for design optimization

device against design constraints and to further tune the device design for improved performance.

9.1.1 Design Flow Using MEMS+

Figure 9.2 illustrates the design flow for device geometry optimization to achieve a specified objective such as maximizing the quality factor, Q, or the output power of the EH. Process awareness is provided by the technology cross section of the EH platform, which includes not only the materials and thicknesses of the various layers but also the cavities below and above the EH element. A parameterized MEMS+ model provides the evaluation of the optimization loop [2] that accounts for the physics of the piezoelectromechanical behavior and for the fluidic damping due to residual gases in the device cavities. Analytical models [3] can be used to estimate the resonant frequencies, Q factors, and beam deflections which can be easily derived in the case of simpler device geometries. We intend to expand the library of the available analytical models to more complex geometries. One important geometric constraint is given by the device cavity depth below and above the harvesting element. Maximum beam deflection is computed using harmonic analysis and verified to satisfy the cavity constraints. The optimization algorithm uses derivative-free techniques, such as genetic programming and simulated annealing, due to their speed and their ability to avoid local optima.

Table 9.1 Optimization strategy. Design space x includes the device geometry from the layout and process stack

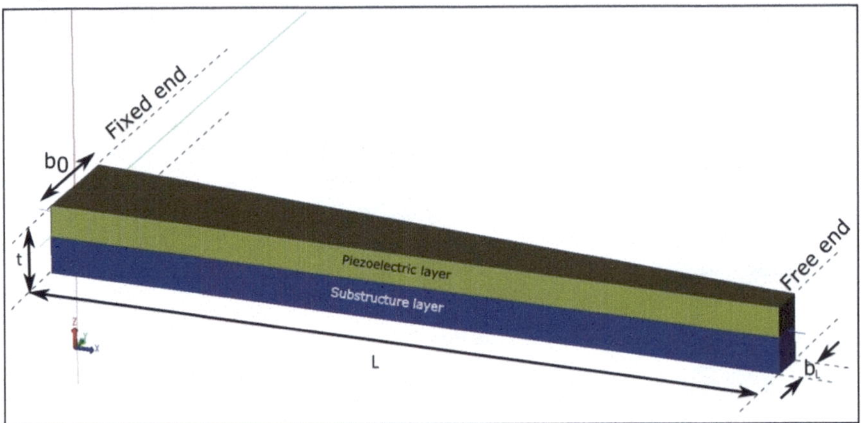

Component	Quantity	Source
Objective: $f(x)$ Maximize	Power density: $f(x) = \frac{P(x)}{A}$	Charge ($q(x)$): MEMS+.
		Device impedance ($Z(x)$): MEMS+.
		Power ($P(x)$): MATLAB model.
		Area (A): Device layout.
Boundaries: Ω	Beam length (l)	Layout template and design rules.
	Proof-mass dimensions, if exists	Layout template and design rules.
	Device width (b_0 and b_1)	Layout template and design rules.
Constraints: $\Phi(x)$	Frequency range	Operating frequency specification.
	Maximum deflection allowed	Cavity depth from process stack.
	Q factor	Bandwidth specification.

9.1.1.1 Device Geometry Optimization

The optimization problem is shown in Table 9.1, where the objective is to maximize power density under interval (linear) and non-linear constraints. The optimization strategy is given as

$$x_{opt} = \arg\max_{x \in \Omega} f(x)\, s.t.\, \Phi(x),$$

where $f(x)$ is the objective function being maximized, Ω is the solution space and is defined by the imposed boundaries, and $\Phi(x)$ is the set of constraints. The physical design rules impose the solution space boundaries and include lower bounds due to critical dimensions and upper bounds due to foundry-imposed die size and rules to ascertain structural integrity such as silicon release and wafer bonding rules. The nonlinear constraints are imposed by technology requirements on cavity depth or design requirements on bandwidth and losses. When the MEMS+ closed optimization loop converges, the optimal design is further verified by running

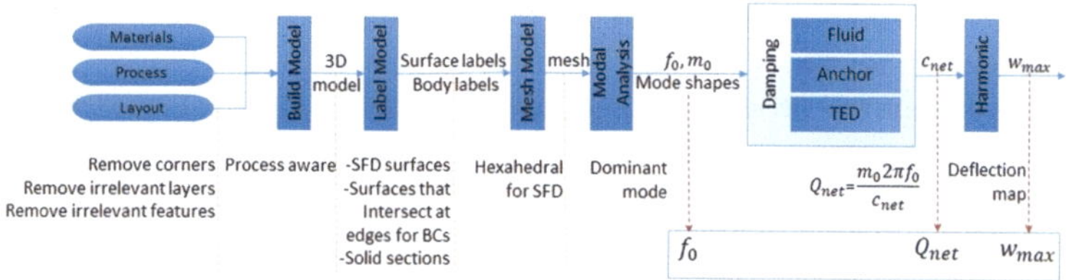

Fig. 9.3 Mechanical modeling using CoventorWare including damping

process-aware 3D FEM analysis using CoventorWare. Once the design is verified, MEMS+ is used to generate its GDSII file for submission to fabrication.

9.1.2 Design Validation Using CoventorWare

CoventorWare helps construct process-aware models of MEMS devices attempting to closely match the real-world device. It is an integrated suite of design and simulation software that constitutes many modeling and simulation features for various MEMS devices. The accompanying field solvers cover multi-physics models such as structural mechanics, coupled electromechanics, electrostatics, piezoelectric, piezoresistive, and damping models. CoventorWare is used for device simulation and analysis using a suite of FEM and BEM solvers. It works by constructing the 3D device model starting with material properties, process flow, and device layout, applying the mesh using one of a list of available mesh algorithms, setting up the multi-physics system, and running simulations before performing data analysis.

CoventorWare helps validate some of MEMS+ models. It is also used to complement the MEMS+ in providing models and data that are not available in MEMS+. The primary use of CoventorWare in the MEMS device design flow employed at MI is to provide a reliable damping model for the Q-factor validation of the candidate designs.

Figure 9.3 describes the CoventorWare workflow for obtaining values for some design targets, namely, the dominant mode resonant frequency, f_0; the maximum beam deflection $w_{\max}$; and the net quality factor including damping computations from the various damping models, Q_{net}, evaluated as a function of the net damping coefficient c_{net} and the effective mass of the structure m_0 at f_0.

9.1.3 Process-Aware Analytical Model

While the FEA-based models prepared using MEMS+ and CoventorWare are more readily available for arbitrary device geometry, analytical models where available can be extremely useful for the following reasons:

1. Analytical solutions to the model are fast to compute thus enabling exhaustive optimal design search.
2. Analytical solutions provide insight into the device dynamics thus helping more informed design decisions.
3. The faster analytical solutions allow the implementation of Monte Carlo simulation and other UQ algorithms to be integrated with the design flow thus making it variability aware.
4. Analytical solutions provide closed-form expressions of the figures of merit of the device or of the governing system of the device for efficient variability analysis using polynomial chaos expansion.

For a straight cantilever PEH we use the model devised by Erturk et al. [4] and for the doubly clamped multielectrode PEH, we use the model developed by Bojesomo et al. [5]. For a tapered PEH, we use the semi-analytical TPEH model derived in Chaps. 4 and 6 for simulation. Analytical damping models [6] can also be employed with these models where possible. The damping coefficients are derived from the results of the damping solver of CoventorWare and plugged into the semi-analytical model. The model is made aware of the process by encoding the layered structure with the appropriate material properties in the model. The multi-layered geometry resulting from the foundry process is modeled by a homogenized bending stiffness YI of the layered structure obtained by adding the bending stiffnesses YI_i of the individual layers computed around the bending axis, $z_{neutral}$, which is given as [7]

$$z_{neutral} = \frac{\sum_{i=1}^{n} Y_i t_i h_i}{\sum_{i=1}^{n} Y_i t_i}.$$

The moment generated at a distance x from the fixed end is then computed as

$$M(x,t) = \sum_{i=1}^{N} - \int_{z_{i-1}-z_{neutral}}^{z_i - z_{neutral}} T_i bz dz = YI \frac{\partial^2 w_{rel}(x,t)}{\partial x},$$

where b is the width of the beam.

9.1.3.1 Adapting ECM Construction to the Device Taper

In the case of the tapered PEH model proposed in Chaps. 4–6, there is an opportunity to speed up the modal analysis and consequently ECM construction by selecting the smallest number of perturbation terms that result in the desired accuracy of modal results for a specific device taper ratio. Figures 5.9 and 5.13 suggest that the smaller the taper ratio is, the fewer perturbation terms are necessary for achieving the desired modal analysis accuracy. In line with this observation, we can scale the number of perturbation terms used with the taper ratio when sweeping the beam taper angles for optimization purposes.

9.2 Motivation of Variability-Aware MEMS Design

As the fabrication processes target smaller features, variability in material and process steps becomes increasingly significant in its impact on device behavior. The consequent increased variability in MEMS process technology requires added margins in the design to guarantee the desired yield. Variation in the fabricated device gets introduced from limitations in process techniques such as inaccurate etching rate, mask misalignments, erroneous lithographic treatment, etc. This process variability causes unexpected variations in the device properties and dimensions like variations in the line width, beam width, and thickness of different layers, as well as properties of deposited and processed material. The variability in device features translates to uncertainty in device performance and consequently a drop in its reliability. Uncertainty quantification (UQ) is particularly important for MEMS where high micromachining tolerances reduce confidence in achieving design specifications. Whereas controlling the tolerances is not in the hands of the designer, modeling the uncertainties and considering them in the design process can help raise confidence in achieving the desired specifications. Uncertainty in the context of the reliability of MEMS is either related to the lack of understanding of the underlying models (epistemic uncertainty) or the variability in fabrication processes and layout (aleatoric uncertainty). These uncertainties are among the biggest challenges to the reliability of MEMS. A design process that has reliability targets must take into account the influence of the input uncertainty on the variability of the device performance.

9.3 Variability-Aware MEMS Device Design Flow

9.3.1 *Design Flow with Variability Constraints*

We build on the general MEMS design guidelines presented by [8] and propose iterating the following steps for process variability-aware MEMS design workflow:

1. Enumerate specifications, performance criteria, design rules, device boundaries, and the reliability target to set up the optimization problem.
2. Define device failure criteria such as the operating frequency being far from the target resonance frequency by a certain margin.
3. Select a reliability level such as a probability of failure less than 5%.
4. Determine the process tolerances and uncertainties to obtain the low-level uncertainty model.
5. Prepare an initial candidate design that satisfies nominal boundaries and constraints and optimizes the cost.
6. Develop the ECM model of the device.
7. Run MC simulation to compute the stochastic model of the ECM.

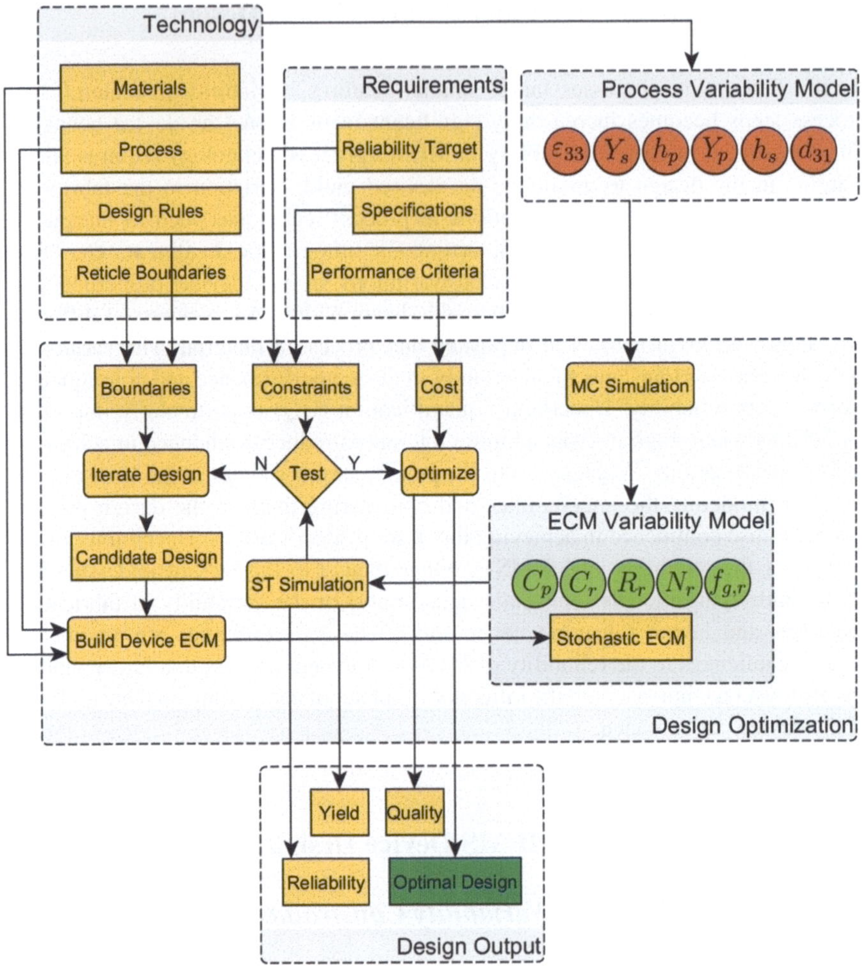

Fig. 9.4 Process variability-aware MEMS design flow with stochastic simulation as part of the optimization loop

8. Run stochastic ECM using ST simulation.
9. Run the forward uncertainty propagation and determine yield and reliability.
10. Select a candidate model that provides the best trade-off between reliability and device performance.

The design workflow is captured in Fig. 9.4.

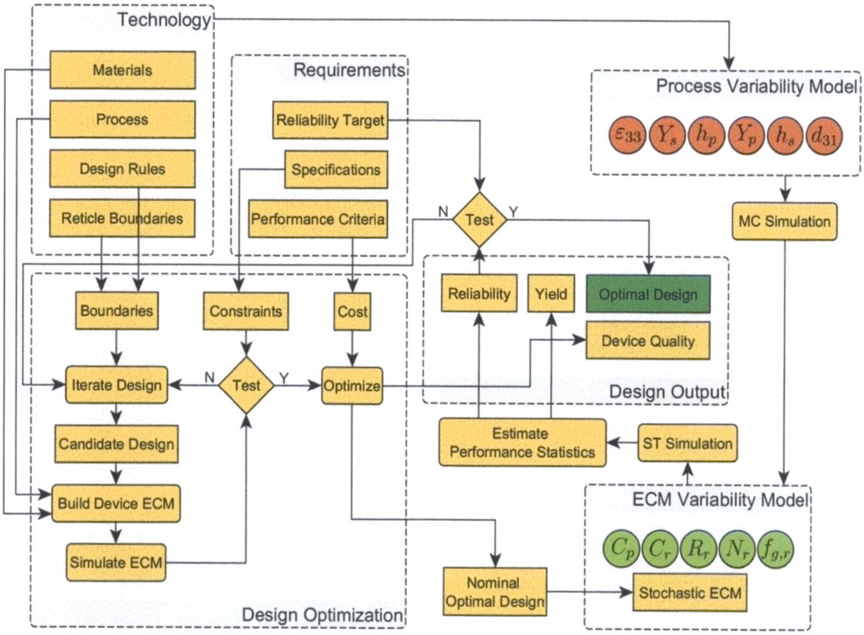

Fig. 9.5 Process variability-aware MEMS design flow with stochastic simulation decoupled from optimization loop

9.3.2 Design Flow with Variability Tests Decoupled

The variability-aware MEMS design workflow shown in Fig. 9.4 integrates the ECM variability model inside of the optimization loop. The optimization over the design space is heavily taxed by stochastic solvers, and to alleviate the computational challenges, we decouple the stochastic simulation and apply it to an optimized candidate design prepared for nominal design specifications. The resulting design flow is shown in Fig. 9.5.

References

1. Yaowen Yang and Lihua Tang. "Equivalent circuit modeling of piezoelectric energy harvesters". In: *Journal of Intelligent Material Systems and Structures* (2009), pp. 2223–2235.
2. *MEMS+ Users Guide*. Coventor Inc. 2002.
3. Alper Erturk and Daniel J Inman. "Issues in mathematical modeling of piezo-electric energy harvesters". In: *Smart Materials and Structures* 17.6 (2008), p. 065016.
4. Alper Erturk and Daniel J Inman. "A distributed parameter electromechanical model for cantilevered piezoelectric energy harvesters". In: *Journal of vibration and acoustics* 130.4 (2008), p. 041002.

5. Alabi Bojesomo, Wajih Syed, and Ibrahim Abe M Elfadel. "Multi-electrode piezoelectric energy harvesters". In: *Circuits and Systems (MWSCAS), 2016 IEEE 59th International Midwest Symposium on*. IEEE. 2016, pp. 1–4.
6. Minhang Bao. *Analysis and design principles of MEMS devices*. Elsevier, 2005.
7. AN Vieira da Silva. "Analytical modeling of the stress-strain distribution in a multilayer structure with applied bending". In: *Master's thesis, Instituto Superior Técnico* (2010).
8. Jonathan W Wittwer, Troy Gomm, and Larry L Howell. "Surface micromachined force gauges: uncertainty and reliability". In: *Journal of Micromechanics and Microengineering* 12.1 (2001), p. 13.

Epilogue

حُسنُ السؤال نصفُ العلم
حديث شريف

Dear Reader, As parting comments, please allow us to highlight three of the unique aspects of this monograph. The first is that it clearly illustrates what academia can do above and beyond the requirements of customer-facing, sponsored research engagements. Indeed, the book accounts for an entirely original approach to model the dynamic behavior of a generic element of MEMS designs, namely, the tapered beam. Second, the modeling approach adopted in this book is fundamental, physics-based, and analytical. It gives primacy to symbolic processing over numerical analysis and to closed-form expressions over computational representations. Third, it has the refreshing contrarian feature of interrogating nature rather than Artificial Intelligence and following physics rather than regression.

In relation to its contrarian feature, one may argue that the symbolic modeling approach of this book would not have been possible without the MATHEMATICA computer algebra system, which owes its origins to a former era of AI when expert systems were programmed to process logical predicates. This argument holds only to the extent that a computer algebra system, like any expert system, is deductive rather than generative. The generative and creative effort in the modeling method of this book lies not in the deductions of MATHEMATICA but rather in the formulation of the symbolic perturbation analysis and in its elaboration into a fully fledged semi-analytical modeling tool.

The symbolic perturbation analysis advocated in this volume is generic in that it can be extended to other physical domains in addition to microelectromechanical systems. As may be observed, domains as varied as acoustics, photonics, structural mechanics, and fluid dynamics can readily profit from the rigorous methodology of

W. U. Syed, I. (Abe) M. Elfadel, *Tapered Beams in MEMS*,
https://doi.org/10.1007/978-3-031-66391-8

this book. From the whisker of a cat to the waveguide of a photonic link and from the trunk of a tree to the rotor blade of a turbofan engine, accurate and efficient models that explicitly take tapering into account can be derived using the symbolic perturbation methodology of the present book.

Appendix A
Mathematical Properties of Perturbation Solutions of Tapered-Beam Systems

A.1 Self-Adjoint Operator Helps Evaluate Eigenvalues

The perturbed tapered system equation is

$$\frac{d^4\psi_n(x)}{dx^4} - \mu_0\psi_n(x) = \sum_{j=1}^{n}\left(\frac{2}{l}\left(\frac{x}{l}\right)^{j-1}\frac{d^3\psi_{n-j}(x)}{dx^3} + \mu_j\psi_{n-j}(x)\right). \tag{A.1}$$

The homogeneous equation of the ITPS in Eq. (A.1), which is also the perturbation equation for the nominal case at $n = 0$, is an eigenvalue problem of the form $(\mathcal{D} - \mu_0 I)\,\psi_n(x) = 0$, where $\mathcal{D} = \frac{d^4}{dx^4}$ and μ_0 its eigenvalue and $\psi_0(x)$, a solution to the rectangular beam mode shape equation, its eigenvector. Let $\mathcal{L} = \mathcal{D} - \mu_0 I$, such that Eq. (A.1) can be written as $\mathcal{L}\psi_n(x) = F_n(x)$ where $F_n(x) = \sum_{j=1}^{n}\left(\frac{2}{l}\left(\frac{x}{l}\right)^{j-1}\frac{d^3\psi_{n-j}(x)}{dx^3} + \mu_j\psi_{n-j}(x)\right)$. If we can prove that the operator $\mathcal{L}$ is a self-adjoint or a Hermitian operator, we can exploit the orthogonality property between the null-space of $\mathcal{L}$ and its image, $\ker\mathcal{L} \perp \mathrm{Im}\mathcal{L}$, to derive an expression for μ_n.

A.2 Proof that the Tapered Beam System is Self-Adjoint

We begin by proving that the system $\mathcal{D}$ is a self-adjoint system. To do so we first find the adjoint operator $\mathcal{D}^*$ such that $\langle \mathcal{D}u \mid v\rangle = \langle u \mid \mathcal{D}^* v\rangle$. Here the inner product is defined in the domain of the beam span as $\langle u(x) \mid v(x)\rangle_x = \int_0^l u(x)v(x)\,dx$. Using this definition of the inner product, we then have

W. U. Syed, I. (Abe) M. Elfadel, *Tapered Beams in MEMS*,
https://doi.org/10.1007/978-3-031-66391-8

$$\langle \mathcal{D}u \mid v\rangle_x = \int_0^l v\mathcal{D}u\,dx = \int_0^l v\frac{d^4u}{dx^4}dx\,.$$

Expanding the integral using integration by parts, we get

$$\int_0^l v\frac{d^4u}{dx^4}dx = v\frac{d^3u}{dx^3} - \frac{dv}{dx}\frac{d^2u}{dx^2} + \frac{d^2v}{dx^2}\frac{du}{dx} - \frac{d^3v}{dx^3}u\bigg|_0^l + \int_0^l \frac{d^4v}{dx^4}u dx\,.$$

We utilize the boundary conditions for the perturbation problem described in Eqs. (4.16) and (4.17) to get

$$\langle \mathcal{D}u \mid v\rangle_x = \int_0^l v\frac{d^4u}{dx^4}dx = \int_0^l \frac{d^4v}{dx^4}udx = \left\langle \mathcal{D}^*u \mid v\right\rangle_x\,,$$

establishing that $\mathcal{D}^* = \mathcal{D} = \frac{d^4}{dx^4}$. This proves that the operator $\mathcal{D}$ is *self-adjoint* and, therefore, $\mathcal{L} = \mathcal{D} - \mu_0 I$ is also self-adjoint. Since $\mathcal{L}$ is also bounded, we can infer that $\mathcal{L}$ is Hermitian.

A.3 Self-Adjointedness Implication on $F_n(x)$

Knowing that the operator $\mathcal{L}$ is self-adjoint, and that $\psi_0(x) \in \ker\mathcal{L}$, that is, it is in the null-space of $\mathcal{L}$, in order to solve for the general nth perturbation order eigenvalue μ_n, we derive the necessary condition for the existence of a solution $\psi_n(x)$ of the equation [1]

$$\mathcal{L}\psi_n(x) = F_n(x)\,,$$

as

$$\begin{aligned}\langle\psi_0(x) \mid F_n(x)\rangle_x &= \langle\psi_0(x) \mid \mathcal{L}\psi_n(x)\rangle_x \\ &= \langle\mathcal{L}\psi_0(x) \mid \psi_n(x)\rangle_x \\ &= \langle 0 \mid \psi_n(x)\rangle_x \\ &= 0\,.\end{aligned}$$

This result confirms that $\psi_0(x)$ is orthogonal to $F_n(x)$, which is the image of $\mathcal{L}$, i.e., $\ker\mathcal{L} \perp \mathrm{Im}\mathcal{L}$.

Appendix B
Some Properties of the Green's Function for Perturbation Solutions of Tapered-Beam Systems

This appendix presents a useful view of the Green's function, namely the eigenvalue expansion view, which highlights the orthogonality property of the Green's function and suggests the need for deflating the system using Fredholm alternative theorem when normalizing the mode shapes.

B.1 Eigenvalue Expansion View of the Green's Function

Take the general system that we are to solve to find the perturbation components given as

$$\mathcal{L}u(x) = \frac{d^4u(x)}{d\xi^4} - \mu_0 u(x) = f(x) .$$

Its related eigenvalue problem can be posed as

$$\mathcal{L}u(x) = -\lambda u(x) , \tag{B.1}$$

where λ is an eigenvalue of $\mathcal{L}$. We can solve Eq. (B.1) by using an eigenfunction expansion of the form

$$u(x) = \sum_{n=1}^{\infty} a_n v_n(x) .$$

Using superposition, Eq. (B.1) can be written as

W. U. Syed, I. (Abe) M. Elfadel, *Tapered Beams in MEMS*,
https://doi.org/10.1007/978-3-031-66391-8

$$\mathcal{L}u(x) = -\sum_{n=1}^{\infty} a_n \lambda_n v_n(x) = f(x) .$$

Taking the inner product of both sides with $v_m(x)$ and using orthogonality, we can show that

$$a_n = -\frac{\int_0^L f(x) v_n(x) dx}{\lambda_n \int_0^L v_n^2(x) dx} .$$

The solution can then be obtained as

$$u(x) = \int_0^L f(\xi) \left(\sum_{n=1}^{\infty} -\frac{v_n(x) v_n(\xi)}{\lambda_n \int_0^L v_n^2(\xi_1) d\xi_1} \right) d\xi .$$

We notice that this equation of the form of finding the system's solution through an inner product with the Green's function as

$$u(x) = \int_0^L f(\xi) G(x, \xi) d\xi .$$

The comparison reveals an eigen-expansion view of the Green's function [2] shown in Eq. (B.2) as

$$G(x, \xi) = \sum_{n=1}^{\infty} -\frac{v_n(x) v_n(\xi)}{\lambda_n \int_0^L v_n^2(\xi_1) d\xi_1} . \tag{B.2}$$

B.2 Fredholm Alternative Theorem for a Boundary Value Problem

Let us consider a BVP consisting of an ODE

$$\mathcal{L}u(x) = f(x) ,$$

subject to homogeneous boundary conditions $\mathcal{B}$, and let us also consider its associated homogeneous adjoint problem

$$\mathcal{L}^* v(x) = 0 ,$$

with associated homogeneous boundary conditions $\mathcal{B}^*$. Then either of the following two statements is true [2]:

1. The solution to the inhomogeneous problem $u(x)$ exists and is unique.
2. At least one nontrivial solution to the homogeneous adjoint problem $v(x) \neq 0$ exists.

If the problem is self-adjoint, that is $\mathcal{L}^* = \mathcal{L}$ and $\mathcal{B}^* = \mathcal{B}$, the adjoint homogeneous problem is simply the homogeneous version of the original BVP.

References

1. John K Hunter. "Asymptotic analysis and singular perturbation theory". In: *Department of Mathematics, University of California at Davis* (2004).
2. William J. Parnell. "Greens functions, integral equations and applications". University of Manchester. 2013.

Index

W. U. Syed, I. (Abe) M. Elfadel, *Tapered Beams in MEMS*,
https://doi.org/10.1007/978-3-031-66391-8

The manufacturer's authorised representative in the EU is Springer Nature Customer Service Centre GmbH, Europaplatz 3, 69115 Heidelberg, Germany. If you have any concerns regarding our products, please contact ProductSafety@springernature.com

Printed and bound by CPI Group (UK) Ltd, Croydon, CR0 4YY
31/07/2026
02182981-0001